Protecting Rice Grains in the Post-Genomic Era

Edited by Yulin Jia

Published in London, United Kingdom

IntechOpen

Supporting open minds since 2005

Protecting Rice Grains in the Post-Genomic Era
http://dx.doi.org/10.5772/intechopen.77851
Edited by Yulin Jia

Contributors
Luis Galán, David Fernández, Jesica Ramírez, Wen-Ming Wang, Jing Fan, John Martin Jerome Jeyakumar, Xin-Gen Zhou, Christian De Guzman, James Oard, Yulin Jia, Melissa Jia, Xueyan Wang, Haijun Zhao, Dr Raimukaram, Noor Sabah, Junjie Xing, Huafeng Deng, Longping Yuan, Chunda Feng, James Correll, Abdul Aziz, Tariq Mahmood, Frank White, Arun Antonyraj

First published in London, United Kingdom, 2019 by IntechOpen
IntechOpen is the global imprint of INTECHOPEN LIMITED, registered in England and Wales, registration number: 11086078, The Shard, 25th floor, 32 London Bridge Street
London, SE19SG – United Kingdom
Printed in Croatia

British Library Cataloguing-in-Publication Data
A catalogue record for this book is available from the British Library

Additional hard and PDF copies can be obtained from orders@intechopen.com

Protecting Rice Grains in the Post-Genomic Era
Edited by Yulin Jia
p. cm.
Print ISBN 978-1-78984-387-3
Online ISBN 978-1-78984-388-0
eBook (PDF) ISBN 978-1-83962-227-4

For EU product safety concerns:
IN TECH d.o.o., Prolaz Marije Krucifikse Kozulić 3, 51000 Rijeka, Croatia, info@intechopen.com or visit our website at intechopen.com.

Meet the editor

Dr. Yulin Jia is a research plant pathologist of USDA Agriculture Research Service Dale Bumpers National Rice Research Center. He received his MS (1993) from the University of Florida and his PhD (1997) from Purdue University. He has been identifying emerging research frontiers, conceiving new objectives, formulating innovative and interdisciplinary approaches, and executing research to help the US rice industry remain competitive in the global marketplace by assuring pest resistance. He has released five mapping populations, and one mutant population consisting of over 20,000 individuals. He has contributed more than 300 publications, including 17 book chapters and 115 peer-reviewed journal articles, including articles in *Nature Genetics*, *Nature Communications*, and *Science*. Dr. Jia is sought out internationally as a mentor and has trained 75 scientists/staff members, including 18 postdoctoral associates and visiting scientists from 10 countries.

Contents

Foreword

The genomics era has given us major advances in our understanding of the rice plant and its relationship with its biophysical environment. The time is right, then, to examine this understanding and speculate where and how we should proceed to address the remaining problems. Yulin Jia and colleagues present an update on some of the most serious problems that have affected rice production. Our understanding of bacterial blight, blast, and sheath blight has led to significant advances in their management and control. However, much postgenomic work remains to be done.

Because rice production has become more intensive and management options more sophisticated, new challenging disease problems are emerging. The treatment of false smut and bacterial panicle blight will be important resources for all working in rice improvement and management. Both of these diseases have long been considered to be minor, almost nuisance, problems in the field. Their emergence to economic levels of importance is a mystery that must be solved. The potential for the false smut to produce mycotoxins adds urgency to the challenge.

There is a welcome inclusion of a range of abiotic challenges such as plant nutrition and arsenic uptake. These are critical to the sustainability and safety of rice production. Placing these in the context of climate change is very important for readers because society seeks to reduce the climate footprint of agriculture worldwide. There are significant tradeoffs that will have to be confronted and this volume should help inform those charged with making decisions.

There is an interesting addition to the disease and management emphasis in the book: the treatment of the nutritional value of rice. There is little doubt that as consumers become more sophisticated, they will demand more nutritious rice. Likewise, rice with starch composition that yields lower glycemic index will be an important future focus for rice breeders. And food processors will require different starch characteristics as they increase the use of rice in processed foods.

The book's orientation is toward important challenges facing rice production in the United States. There is very good treatment of a number of specific breeding challenges around hybrid rice and indeed unique opportunities offered with hybrid rice breeding. Nonetheless, the authors address a number of areas that are of broad interest to the global rice community. This work is ambitious and is an important resource for students and all those engaged in the production side of the rice sector.

Robert S. Zeigler
Director General Emeritus,
International Rice Research Institute

Preface

Rice (*Oryza sativa*) feeds more than 3.5 billion people on planet Earth, but more rice will be needed to meet the demand of population growth. Rice requires more water compared to other crops, as noted in the translation for the Chinese name for rice, "Water Grain." Today, rice has been grown in a wide range of areas from rainfed low- and uplands to irrigated areas and in many areas where other crops would fail. As the standard of living increases, more people are moving to cities and more rice fields are being replaced due to urban development. Rice production has been subjected to an increasing amount of constraints, including shortages of arable land, labor, water, and biotic and abiotic stressors. Consequentially, the supply of water is decreasing drastically: more wetlands will disappear, and more water will be contaminated, resulting in an increased concern for sustainability and the environmental consequences of rice production.

Rice was the first crop whose genome sequence was reliably determined and for whom tremendous resources composed of worldwide rice germplasm collections was assembled. Many mapping populations, genetic stocks, mutants and wild rice relatives, and high-density genetic and expression maps for rice have also been accumulated rapidly. Many of the global research accomplishments during the early 21st century have built the fundamental basis of rice grain protection through an improved understanding of genetics, genomics, and host–pathogen interactions.

In this book, a wide range of topics is included such as rice germplasm collection, methods to increase yield, nutritional benefits, rice response mechanisms to increasingly virulent pathogens, functional genomics, and host–pathogen interactions. This book is composed of 12 chapters from scientists around the globe and it is hoped that it will spread their ideas for and successes with safeguarding rice crops to ensure sufficient food supplies that meet human needs on Earth for years to come.

I hope that the knowledge from this book will benefit rice specialists, farmers, and students who are interested in molecular genetics, biology, pathology, and crop protection. This book could not have been prepared without the help of many scientists and specialists who generously contributed their time and expertise in writing and production. All authors of this book are appreciated for their extra efforts to contribute their knowledge, skills, visions, and philosophies despite the various imposing deadlines. Special appreciation is given to my daughter Mary S. Jia who assisted me in editing and completing this book. The top-notch research that has been conducted around the world is fragmented and unhelpful to students and other scientists who are seeking rice crop improvement. It is only through comprehensive reviews, such as this book, that the spread of such knowledge may be accomplished to propagate the current techniques and theories that may, one day, facilitate world peace.

This book is dedicated to Professor Wenrong Liu (a plant pathology laboratory instructor of Xichang Agricultural College, China), deceased on February 20, 1992, and Mr. Robert Spencer Edsall Jr. (a farmer, USA), deceased on May 10, 2019, who were Dr. Jia's former mentors on plant pathology and farming practice. Mr. Edsall was also one of the sponsors for Dr. Jia's MS (1993) of the University of Florida through Edsall Groves Service Inc., Florida, USA, and for Communicating for Agricultural Exchange Program.

Yulin Jia
US Department of Agriculture,
Agricultural Research Service,
Dale Bumpers National Rice Research Center,
USA

Section 1

Resistance Resources, Pathogen Genetics, Host and Pathogen Interaction, Disease Management

Chapter 1

Introductory Chapter: Protecting Rice Grains in the Post-Genomic Era: Are We There Yet?

Yulin Jia

1. Introduction

Rice (*Oryza sativa* L.) was domesticated roughly 10,000–14,000 years ago. Today, rice is the most widely grown food crop in approximately 113 countries, providing a major dietary caloric and protein supply (**Figure 1**).

Rice has been a staple food source for many cultures globally and is particularly critical for developing countries where other food resources are limited. Rice

Figure 1.
Pictures showing rice plants grown in the USA, 2018. Blast-resistant rice breeding and differential lines grown in Crowley, Louisiana (a, b). Blast-resistant plants and two mapping populations with blast resistance genes grown in a greenhouse and a field, respectively, Stuttgart, Arkansas, USA (c, d). Pictures were taken from rice grown in experimental stations in the USA with an iPhone.

IntechOpen

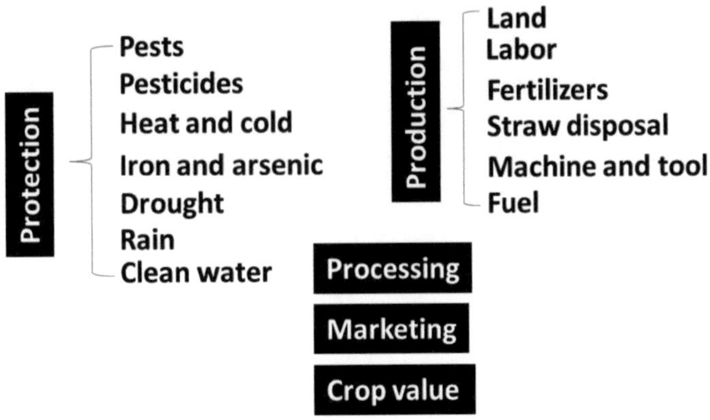

Figure 2.
Diagram showing the major challenges of rice protection.

grains feed more than 3.5 billion people [1]. Rice germinates under a wide range of temperatures, from 10 to 40°C, and requires about 3–6 months to mature. There are two primary methods of rice production: transplanting of rice seedlings from a nursery to paddy rice fields to ensure uniformity and weed control and sowing seeds in a rice field directly to conserve labor. Rice production was particularly enhanced during the green revolution through the deployment of high-yielding uniform rice varieties, especially semidwarf rice. However, the wide deployment of well-adapted uniform varieties consequently made rice more vulnerable to both biotic and abiotic stressors (**Figure 2**). Many major rice-growing countries today including China, India, Japan, Korea, Brazil, the USA, the Philippines, and Thailand have been experiencing old and new obstacles that inhibit the stable production of rice. As such, the question of whether the current methods of rice cultivation are sufficient, or whether we are there yet, or have plateaued, remains unanswered.

2. Major challenges and opportunities for rice grains

Major challenges and opportunities for rice grains are embedded during rice propagation, protection, postharvest pest management, processing, and marketing (**Figure 2**). First, rice diseases have had a major impact on the stability of rice production. For example, the famine in Bengal, India, in 1942 was mainly due to rice brown spot disease [2], and a rice blast epidemic in Korea caused a major food shortage in the 1970s [3]. The most common rice diseases are caused by fungal and bacterial pathogens. Fungal diseases include rice blast, sheath blight, false smut, rice brown spot, rice stackburn, leaf smut (**Figure 3**), and grain discoloration diseases include rice blast, bacterial panicle blight, false smut, and kernel smut (**Figure 4**).

Bacterial diseases include bacterial blight and bacterial panicle blight (**Figure 4c**). Insects are another type of stressor for rice production, whose importance varies with the country, location, and time. Rice water weevil (*Lissorhoptrus oryzophilus*), rice stink bug (*Oebalus pugnax*), fall armyworm (*Spodoptera frugiperda*), rice stalk borer (*Chilo plejadellus*), and grasshoppers (*Orthoptera*) are commonly found in the southern USA. For decades, the common remedy for pest management has been an increased application of pesticides and the alteration of cultural practices with limited use of natural genetic resistance. Most rice pathogens and insects adapt to their food sources through genetic changes, resulting

Figure 3.
Photographs showing examples of rice diseases on leaf, sheath, and grains. (a) Leaf blast caused by M. oryzae; (b) sheath blight caused by Rhizoctonia solani; (c) brown spot caused by Cochliobolus miyabeanus, called Helminthosporium leaf spot; (d) stackburn caused by Alternaria padwickii; (e) rice leaf smut caused by Entyloma oryzae; and (f) straighthead disorder. Pictures were taken from diseased plants under greenhouse and field conditions in Stuttgart, Arkansas, USA.

in more virulent races. Consequently, rice pests are becoming more resistant to pesticides and newly deployed resistance genes.

Rice thrives in a wide range of geographic and climatic regions, especially where many other crops would fail. As a semiaquatic crop, rice uniquely requires a lot of water as evident by its Chinese name, "water grain." A lack of sufficient clean water is problematic for producing healthy and safe rice as the presence of water is particularly critical during the rice vegetative and reproductive growth stages. There are numerous reasons for the current limit in clean water supply for rice production: depletion of groundwater, increased water usage by human consumption, and contamination due to improper disposal of industrial waste and pesticide

Figure 4.
Examples of rice diseases under field and greenhouse conditions. (a) Rice blast disease showing leaf and kernel blast; (b) sheath blight disease showing symptoms on leaf and stems; (c) bacterial panicle blight disease in a field (3 days after detached leaves from five susceptible rice varieties were inoculated with B. glumae), top; (d) rice false smut disease with two 2× enlarged detached smut balls in the top; (e) rice false smut disease in a field showing the late stage of false smut balls; (f) rice kernel smut with two 3× enlarged smut kernels on the top (all pictures were based on predicted symptoms of rice in greenhouses and fields in Crowley, Louisiana, and Stuttgart, Arkansas, USA).

contamination [4]. The abiotic stressors of rice include extreme temperatures during the seedling and reproductive stages; high concentrations of salt; soil heavy metals including cadmium (Cd), lead (Pb), and arsenic (As); and drought. Straighthead is the most common abiotic disorder in the USA (**Figure 3f**). However, the importance of abiotic stress varies with the region in which the rice is being grown.

It is evident that human life in the twenty-first century is much better than that of past centuries due to continued economic growth. As more people are being liberated from starvation, a demand for improved infrastructure to accommodate the incoming population has increased [5]. However, global urban land expansion often encroaches on croplands necessary for rice production. It was predicted that urban expansion will result in the loss of approximately 1.8–2.4% of croplands by 2030 [6]. Most of these land losses take place in the productive lands in top rice-producing countries in Asia. As a result, the lost croplands were predicted to be responsible for 3–4% of worldwide crop production loss in 2000 [6].

Despite rice being a critical food crop for humanity, the monetary value of rice is extremely low. Farmers must grow rice in the interest of the continued survival of humanity, but increasing the crop value could cause an instability in food security and, as a result, society as well. Therefore, rice is not a competitive crop compared to many other nonessential agricultural products that generate higher income. Rice investment, as such, is among the lowest in many parts of the world. The current obvious challenge is developing methods of protecting rice grains despite its insufficient funding and resources. Today, the majority of rice is grown by small-scale farmers for local consumption [7]. The global market is less than 10% of total rice production with unique restrictions on the trade. Accordingly, the rice market is volatile and distorted [8]. Clearly, it is safe to conclude that human intervention is the most impacting factor that directly influences the outcomes of the interactions of rice with pests and with environmental conditions (**Figure 5**).

The mission of this book is not to solve the problems of yesterday but to solve those of today and tomorrow. It was estimated that rice yield must increase by more than 1.2% annually to meet the demand for food security as of 2005 [9]. The ability to sustainably grow rice is essential for the continued growth of the human population on the planet Earth. However, a shortage of labor will drive the adoption of direct seeding rather than transplanting and will impact the uniformity of space for each individual plant creating a microclimate conducive to disease epidemics. Additionally, water shortage will not only reduce the productive growth of rice plants but will also result in more crop losses caused by rice blast disease [10]. It is of utmost importance that each of us, everywhere, increases the efficiency of developing long-lasting resistance to stressors. In order to achieve this goal, a better understanding of the available scientific knowledge of rice host-pathogen interactions and its optimal environment (climate and human resources, marketing and rice utilization) will be needed.

Rice was the first food crop whose genome sequence was determined [11–13]. Shortly after, high-resolution genetic maps, expression maps, databases, and bioinformatics tools for rice research were developed [14–17]. These genomic resources were used to develop a basic molecular toolbox for rice breeding and crop

Figure 5.
Graphic presentation of primary factors that impact the rice bowl. The human intervention of host-pathogen-environment interactions is the key factor for securing the rice bowl. The Rice bowl at the center was taken with an iPhone.

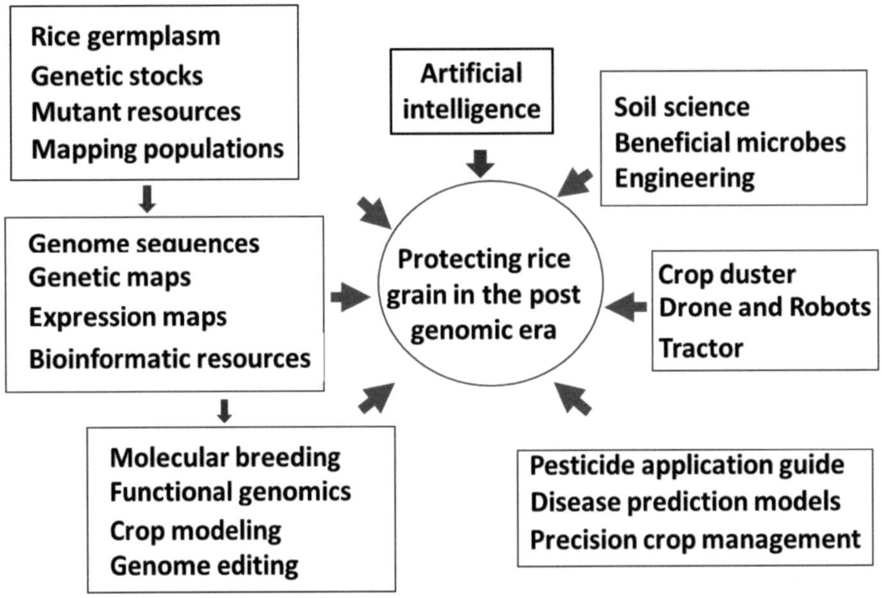

Figure 6.
Diagram showing tools and resources for protecting rice grains in the post-genomic era.

protection. Most recently, advances with next-generation DNA sequencing, coupled with rapidly developed bioinformatics tools, have paved a road for a deeper understanding of genetics and genomics of rice, pathogens, and host-pathogen interactions under changing environmental conditions [18, 19] (**Figure 6**).

The majority of chapters in this book describe the current update on plant resources, genomics, and methods for crop protection and production. This book also includes the methods to improve the market value of rice and better cultural practices and rice processing to ensure profits and marketability of rice grains. These tools, resources, and knowledge are implemented interchangeably into the current efforts to produce rice sustainably (**Figure 5**).

3. Conclusion

The book presents the power of genetics, genomics, and modern cultural practices for the production of one of the most important food crops, rice. The strategies and knowledge extracted from various rice resources, discovered resistance genes, observed mechanisms of host-pathogen interactions, cultural practices of rice labor, and rice processing and utilization of rice grains and husk are highlighted. It is anticipated that the knowledge in this book will guide stable rice production, protection, processing, and marketing. This book may also be useful for students and specialists who are interested in plant pathology, genetics, molecular biology, physiology, agronomy, and biological engineering.

Acknowledgements

The author thanks Mary Jia of Arkansas School for Mathematics, Sciences and the Arts, Hot Springs, AR, USA; Heather Box, Alan Sites, Nick Herrings, and Melissa H. Jia and other staff members of the USDA ARS DB NRRC, Stuttgart, AR, USA; and Dr. Yeshi Wamishe and Md Mamunur Rasheed and other staff members of the University of Arkansas Rice Research and Extension Center, Stuttgart, Arkansas, for excellent technical assistance and useful discussions. For critical reviews the author thanks Drs. Jai Rohila and Trevis Huggins of the USDA ARS DB NRRC and Yong-Bao Pan of the USDA ARS, Houma, LA, USA. Mention of a trademark or proprietary product does not constitute a guarantee or warranty of the product by the US Department of Agriculture and does not imply its approval to the exclusion of other products that also can be suitable. The USDA is an equal opportunity provider and employer. All experiments complied with the current laws of the USA, the country in which they were performed.

Author details

Yulin Jia
USDA-ARS, Dale Bumpers National Rice Research Center, Stuttgart, AR, USA

*Address all correspondence to: yulin.jia@ars.usda.gov

References

[1] International Rice Research Institute. World Rice Statistics 2013. Los Baños, Philippines: IRRI; 2013. Available from: http://irri.org/ [Accessed: November 27, 2018]

[2] Padmanabhan SY. The great Bengal famine. Annual Review of Phytopathology. 1973;**11**:11-26

[3] Ou SH. Rice Diseases. 2nd ed. Kew, Surrey, England: Commonwealth Mycological Institute; 1985

[4] Jiang J, Wu F, Liu Y, Deng X. Modeling the impacts of urbanization and industrial transformation on water resources in China: An integrated hydro-economic CGE analysis. Sustainability. 2016;**6**:7586-7600. DOI: 10.3390/su6117586

[5] Okamoto K, Sharifi A, Chiba Y. The impact of urbanization on land use and the hanging role of forests in Vientiane. In: Yokoyama S et al., editors. Integrated Studies of Social and Natural Environmental Transition in Laos, Advances in Asian Human-Environmental Research. Japan: Springer; 2014. pp. 29-38. DOI: 10.1007/978-4-431-54956-7_2

[6] Bren d, Amour C, Reitsma F, Baiocchi G, Barthel S, Guneralp B, Erb K-H, et al. Future urban land expansion and implications for global croplands. Proceedings of the National Academy of Sciences of the United States. 2017;**114**(34):8939-8944

[7] Rice is life: Increased, Sustainable Rice Production Key to Global Food Security. Available from: http://www.fao.org/newsroom/EN/focus/2004/36887/index.html [Accessed: November 27, 2018]

[8] Muthayya S, Sugimoto JD, Montgomery S, Maberly GF. An overview of global rice production, supply, trade and consumption. Annals of the New York Academy of Sciences. 2014;**1324**:7-14. DOI: 10.1111/nyas.12540

[9] Normile D. Reinventing rice to feed the world. Science. 2008;**321**:330-333

[10] Lai XH, Marchetti MA, Petersen HD. Comparative slow-blasting in rice grown under upland and flooded blast nursery culture. Plant Disease. 1999;**83**:681-684

[11] Goff SA, Ricke D, Lan TH, Presting G, Wang R, Dunn M, et al. A draft sequence of the rice genome (*Oryza sativa* L. ssp. *japonica*). Science. 2002;**296**(5565):92-100. DOI: 10.1126/science.1068275.

[12] Yu J, Hu S, Wang J, Wong GK, Li S, Liu B, et al. A draft sequence of the rice genome (*Oryza sativa* L. ssp. *indica*). Science. 2002;**296**(5565):79-92. DOI: 10.1126/science.1068037

[13] International Rice Genome Sequencing Project. The map-based sequence of the rice genome. Nature. 2005;**436**(7052):793-800. DOI: 10.1038/nature03895

[14] Jung KH, An G, Ronald PC. Towards a better bowl of rice: Assigning function to tens of thousands of rice genes. Nature Reviews. Genetics. 2008;**9**(2):91-101

[15] Feltus FA, Wan J, Schulze SR, Estill JC, Jiang N, Paterson AH. An SNP resource for rice genetics and breeding based on subspecies *indica* and *japonica* genome alignments. Genome Research. 2004;**14**(9):1812-1819. DOI: 10.1101/gr.2479404

[16] Ni J, Pujar A, Youens-Clark K, Yap I, Jaiswal P, Tecle I, et al. Gramene

QTL database: Development, content and applications. Database. 2009;**2009**:bap005

[17] Jackson SA. Rice: The first crop genome. Rice. 2016;**9**:14. DOI: 10.1186/s12284-016-0087-4

[18] Shendure J, Ji H. Next-generation DNA sequencing. Nature Biotechnology. 2008;**26**(10):1135-1145. DOI: 10.1038/nbt1486

[19] Garg P, Jaiswal P. Databases and bioinformatics tools for rice research. Current Plant Biology. 2016;**7-8**:39-52

Chapter 2

Genetics and Breeding System for Cytoplasmic and Genetic Male Sterility in Rice

Christian De Guzman and James Oard

Abstract

The initial discovery of cytoplasmic male sterile (CMS) three-line system made it possible to produce hybrids that significantly increase rice yields compared to its inbred counterparts. Further genetic and molecular studies help elucidate the mechanisms involved in CMS male sterility. Additional CMS types were also discovered with similar genetic control from wild sources by interspecific hybridization. While the three-line system was a success, the two line system using photoperiod genetic male sterile (PGMS), thermosensitive genetic male sterile (TGMS) and photoperiod and thermosensitive genetic male sterile (PTGMS) were becoming more popular due to the ease in breeding and with more hybrid combinations generated compared to the CMS types. Inheritance and molecular studies showed that the trait is controlled by one or more recessive genes depending on the genetic background and environmental conditions. Due to the sensitivity of the lines to temperature and/or photoperiod, unique breeding procedures were followed. Methods involved the use of growth chamber, timing of planting, and selection of suitable locations. These practices successfully maintained sterility for hybrid seed production or reversion to fertility for seed multiplication of parental male sterile lines.

Keywords: cytoplasmic male sterile (CMS), genetic male sterile, hybrid rice, photoperiod, thermosensitive

1. Introduction

The use of hybrid greatly increased rice production worldwide due to the improved yields, better tolerance to pest, diseases and environmental stress compared to inbred varieties. The discovery of cytoplasmic male sterility was the major milestone to the development of hybrid rice [1]. Further discovery of two line male sterility made hybrid breeding more efficient and further increased the probability in finding the best performing hybrid combinations [2].

Two different male sterility systems are available for hybrid seed production (**Figure 1**). The first is a cytoplasmic male sterility (CMS) which is a three-line system that uses a male sterile line, a restorer line and a maintainer line. The male sterility is more stable albeit more complicated to breed and maintain [3]. The second is the two-line male sterility system that uses a genetic male sterile which is controlled by temperature, photoperiod or both. The use of this system is increasing due to the ease in breeding, finding more heterotic combinations, and in seed multiplication of parental male sterile lines. However, hybrid seed production may

 IntechOpen

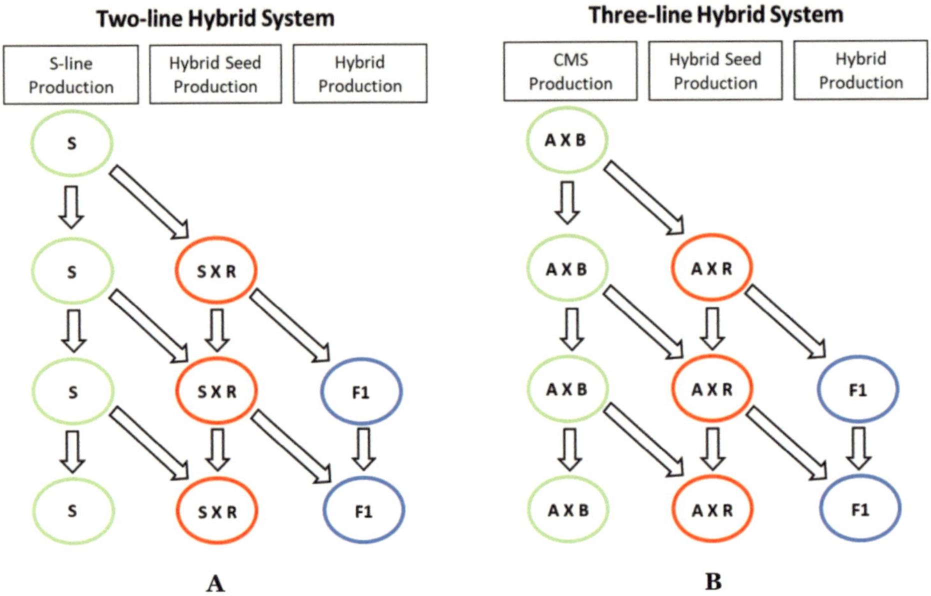

Figure 1.
Comparison of two-line and three-line hybrid rice breeding system. A: two line hybrid system; S—genetic male sterile, R—restorer/pollen fertile, F1—hybrid. B: three line hybrid system A—CMS line, B—maintainer line, R—restorer line with restorer gene, F1—hybrid.

be catastrophic if there are severe changes in environmental conditions [4]. Both systems proved effective in hybrid rice production which increased yields by up to 20% therefore, increased farm profitability and has contributed significantly in addressing global food security.

Heterosis in hybrid rice minimize the impact of reduced yields brought by diseases compared to the inbred counterpart. However, due to narrow genetic diversity of the male sterile parent, they became vulnerable to pathogens and pests resulting to the loss of its yield potential [5]. This makes it difficult for growers to recover the high cost of seed and F_1 production. It became apparent that discovering new sources of male sterility to increase genetic diversity and further introgression of resistance genes are necessary to secure the yield gain in hybrid rice [6].

This chapter focuses on the discovery of rice male sterility, genetics, mechanisms and procedures in multiplication and handling of male sterile rice for hybrid rice breeding.

2. Cytoplasmic male sterile (CMS) rice lines

Development and cultivation of hybrid rice started in China with the initial work of rice breeder Yuan Longping. As early as 1964, Yuan Longping have tested different male sterile lines however, no stable sterility exists and the group started resorting to making distant hybridization by crossing wild rice with cultivated rice. In 1970, a wild-abortive type cytoplasmic male sterile rice CMS-WA were discovered which eventually leads to the release of the first hybrid rice in 1976. By 1980's, hybrid rice accounts to about 55% of the total rice planting area in China [7, 8]. More CMS types were discovered that further expand the diversity in hybrid rice three-line system. These were developed by direct crossing or backcross breeding from two different species, subspecies or different cultivars [9]. The major type of CMS systems with their cytoplasm and nucleus sources are shown in **Table 1**.

MS type	Male sterility source	
	Cytoplasm	Nucleus
CMS-BT	Chinsurah Boro II (*indica*)	Liming (*japonica*)
CMS-HL	Hong lian (*Oryza rufipogon*)	Liantanzao (*indica*)
CMS-CW	Chinese wild W1 (*Oryza rufipogon*)	Reimei (*japonica*)
CMS-WA	Wild abortive (*Oryza rufipogon*)	Erjiunan (*indica*)
CMS-LD	Burmese "Lead rice" (*indica*)	Fujisaka 5 (*japonica*)

Table 1.
Primary CMS male sterility systems utilized in hybrid rice production.

There were more than 60 types of CMS systems discovered in China alone but most of them may only be classified in three types CMS-BT (Boro II), CMS-WA (wild abortive), and CMS-HL (Honglian) [10, 11]. The three major types produces pollen that lack starch or are starch deficient while CMS-LD and CMS-CW produces morphologically normal pollen grains but were unable to fully germinate [12]. In CMS-WA, pollen abortion occur at a uninucleate stage primarily during microspore development [13]. The result is an irregularly shaped and lightly stained pollen when treated with 1% iodine potassium iodide solution (I_2KI). The genotype of sporophytic tissues determines pollen abortion. In CMS-BT, pollen abortion occurs at trinucleate stage with pollen lightly stained due to deficiency of starch and spherical in shape rather than irregular [14]. In CMS-HL, pollen abortion appears at binucleate stage and the pollen is spherical in shape but without starch. Restoration to fertility in all CMS type except CMS-WA are all gametophytic therefore producing half of the pollen fertile in the F_1 generation (**Figure 2**).

2.1 Genetics and mechanism of cytoplasmic male sterility (CMS)

Sterility in CMS is controlled by the interaction of genes in the cytoplasm and the nucleus. The sterility factor S is located in the mitochondrial DNA while the *rf* (restorer of fertility) allele is located in the nucleus. The plant is sterile (A line) if it carries both the S factor and the recessive allele *rf*. Maintainer line (B line) carries the *rf* allele but has a different cytoplasmic factor N which allows the plant to be fertile. The B-line has the ability to make the S line produce seeds after crossing but the progeny remains sterile thus useful for S line seed multiplication. Restorers (R line) is the diverse pollen fertile parent that carries the dominant *Rf* gene that when crossed to B line restores fertility in the F_1 (**Figure 3**) [15].

CMS-BT genes were the first to be identified that has the mitochondrial open reading frame *orf79* [16] and is co-transcribed with *B-atp-orf79*. Similarly, CMS-HL carries the mitochondrial *atp6-orfH79* in which *orfH79* and *orf79* are 98% identical in DNA sequence [9]. In CMS-WA *orf224*, *orf284* and *orf288* were discovered with one still unknown segment. Together they encode a 325-residue protein with three transmembrane segments that are believed to be responsible for CMS trait [17]. A *B-atp-orf79* like structures were also found in CMS-LD that may be link to male sterility but in CMS-CW, no similar structures were identified, thus the specific cytoplasmic factor is still unknown [17].

A total of six restorer of fertility genes (*Rf*) in rice have been discovered. These are *Rf1a*, *Rf1b*, *Rf2*, *Rf4*, *Rf5*, and *Rf17*. These genes were classified as pentatricopeptide repeat (PPR) proteins which are RNA binding and act in post-transcriptional mRNA process in cell organelles [18]. *Rf1a* and *Rf1b* restores fertility in CMS-BT while in CMS-HL, 50% fertile pollen can be restored by either *Rf5* or *Rf6*. If both present in

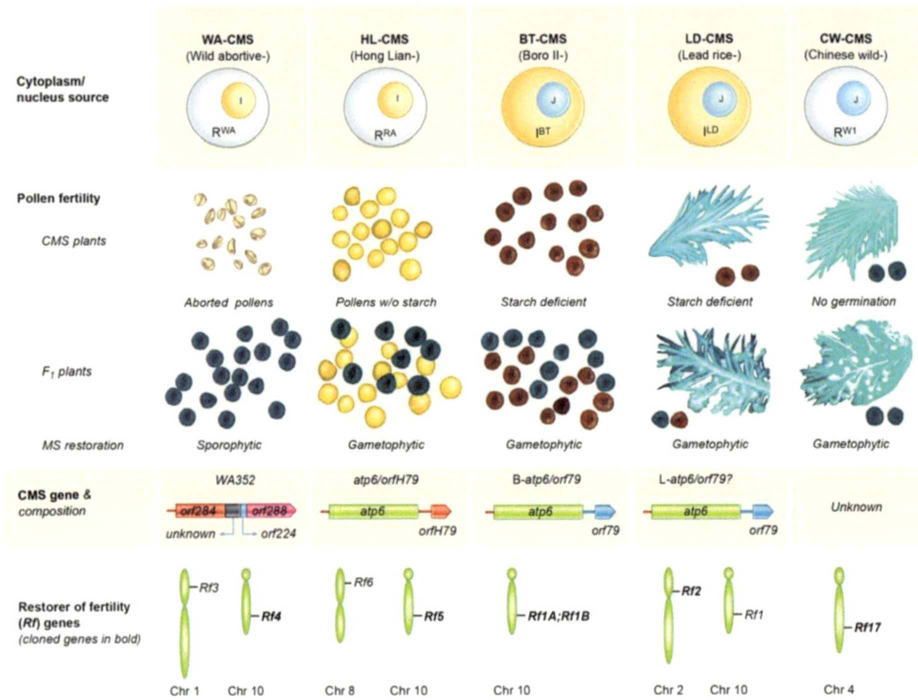

Figure 2.
A schematic presentation of the five well-studied rice CMS types. Abbreviations for cytoplasm sources are RWA for wild-abortive Oryza rufipogon, RRA for red-awned O. rufipogon, and RW1 for Chinese wild rice (O. rufipogon) accession W1; IBT and ILD for indica Boro-II type and Lead rice, respectively. Nucleus sources are either indica (I) or japonica (J) [7].

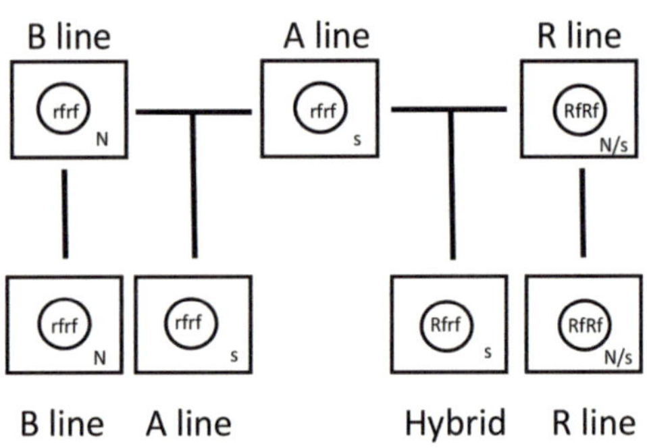

Figure 3.
Schematic diagram of CMS three line system. Rfrf = nuclear gene homozygote recessive, RfRf = nuclear gene homozygote dominant, Rfrf = nuclear gene heterozygous, N = cytoplasmic factor (fertile), s = cytoplasmic factor (sterile).

CMS-HL, *Rf5* and *Rf6* restores 75% of fertile pollen in the F_1. Further analysis by sequencing and cloning concluded that *Rf1* is the same gene as *Rf5* located in chromosome 10 [19]. *Rf3* and *Rf4* restores fertility in CMS-WA and are located in chromosome 1 and 10 respectively with the latter recently cloned [20]. CMS-LD fertility can be restored by *Rf1* or *Rf2* while CMS-CW can only be restored by a single gene *Rf17* [12].

2.2 Breeding and diversification of CMS sources

Diversification of both CMS maintainers and restorer lines are very important to guarantee the continued progress of finding the best hybrid combinations.

Extensive evaluation of lines and backcross breeding were employed to improve the lines and adapt to a particular environment [21]. Furthermore, new maintainers and restorers were developed from the original donors. A new CMS source was discovered in Dongxiang wild rice by continuous backcrossing to the *indica* variety Zongzao 35 [10]. In another study, a new source was found from interspecific cross of an *indica* with an African rice (*Oryza glaberrima Steud.*) that showed similarity to the sporophytic type CMS-WA [22].

3. Discovery of genetic male sterile lines

Although the CMS three-line system greatly increased yields in hybrid rice, there are difficulties and limitations on its use. One of the difficulties is the need to simultaneously develop maintainer lines (B lines) by subsequent nucleus substitution of the original CMS line with the B lines through repeated backcrossing. Furthermore, there are also limited choices available for restorer lines (R lines) with only about 5% of the current existing lines can be used that carries the restorer gene [7]. The discovery of genetic male sterility or photoperiod and/or thermosensitive male sterile lines addresses these problems. These lines responds to photoperiod, temperature or a combination of both which cause the plant to be fertile or sterile depending on the critical daylength or temperature [23]. With the two-line system using genetic male sterile, there is no need to develop a maintainer line and any fertile line can be used as a restorer. This greatly reduce the time and resources in making hybrid combination and parental seed production. Moreover, it broadens the available choices of restorers that can generate more combinations which in turn increases the probability of finding the best hybrids [4].

Extensive studies suggest that genetic male sterile lines can be broadly classified into three categories; photoperiod genetic male sterile (PGMS), thermosensitive genetic male sterile (TGMS), and photoperiod and thermosensitive genetic male sterile (PTGMS) [15, 24, 25]. The first reported genetic male sterile came from spontaneous mutant in a japonica cultivar Nongken 58 discovered in Hubei China and were later called as Nongken 58S [2]. Further studies after its discovery revealed that the male sterility is regulated mainly by photoperiod and thus referred to as photoperiod genetic male sterile (PGMS). Nongken 58S showed complete pollen sterility when grown under long day conditions (>14 h), fertility was restored when subjected to <10 h of light under controlled environment [26].

A thermosensitive type of male sterility was discovered in a spontaneous mutant AnnongS-1 (Ans-1) in 1997. The pollen remained sterile at both long and short day when exposed to 33°C and reverts back to fertile when the temperature reached 24°C [27]. Additional lines exhibiting thermosensitivity were also discovered in Zhu1S, Hengnong 1S and Guangzhang 63S where the fertility rates vary at different controlled temperatures regardless of daylength [28–30].

The third classification of genetic male sterility affects both photoperiod and temperature. Pei'ai 64S is a line derived from the original male sterile mutant Nongken 58S with genetic backgrounds such as *indica* and *javanica* [31]. A study conducted on the response of Pei'ai 64S and another line 8902S showed fertility under long daylength (>14.5 h) and low temperature (24°C) or short daylength (10 h) and high temperature (28°C) conditions, but were consistently sterile at long daylength (14 h) and high temperature conditions (28°C) [32].

3.1 Genetics of male sterility

Numerous genetic studies concluded that genetic male sterility can be controlled by single, two genes or multiple genes depending on the genetic background and the environment. The original Nongken 58S when crossed to conventional *indica* and *japonica* lines produced F_1's that are all fertile [15]. The F_2 reciprocal crosses to Nongken 58S concluded that that male sterility is controlled by a single recessive gene [33]. Further studies showed similar results when Nongken 58S (sterile) and Nongken 58 (fertile) were crossed showing a typical single gene recessive segregation in the F_2 population under long day conditions [34, 35]. A single locus segregation of genetic male sterile lines was also reported in several TGMS lines under long day and high temperature field conditions [29, 30, 36, 37]. Previous research also discovered two gene recessive segregation when Nongken 58S was crossed to an *indica* variety. F_2 segregation exhibited a ratio of 15 fertile: 1 sterile pollen fertility [2]. Several *indica* derived male sterile lines such as Pei'ai 64S displayed similar genetic ratios [31, 38, 39]. Additional studies detected segregation that followed either a bimodal or a continuous distribution in some populations of genetic male steriles [26, 40].

Studies conducted under US conditions on *indica* male sterile line 2008S originated from China showed that male sterility is controlled by two or three recessive genes depending on the location and year of planting (**Figure 4**). When planted in

Figure 4.
Male sterile line 2009S grown in the greenhouse (A) and in the field (C). 2008S male sterile line grown in the greenhouse (B) and in the field (D). Both lines exhibited 100% sterility when grown at H. Rouse Caffey Rice Research Station in Crowley, Louisiana USA under high temperature and long day conditions.

Stuttgart Arkansas USA, 2008SxCL131 F_2 population fit a three-gene model, while the same population planted in Crowley, Louisiana USA segregated in two-gene model during 2013. In 2014, the 2008S/Cypress population fit both 15:1 and 63:1 ratios in Arkansas, whereas 2008S/CL131 only fit a 15:1 segregation ratio. In the same year at Crowley, Louisiana a single 15:1 ratio exhibited in 2008SxCypress, and only a three-gene model was found for 2008SxCL131 [4].

Another male sterile line 2009S currently used by the LSU Agcenter for their hybrid rice breeding program shows a single gene recessive inheritance (**Figure 4**). Field trials carried out in Crowley Louisiana over 2 years (2013–2014) in two F_2 and four BC_1F_2 populations showed consistent results both in seed and pollen fertilities [25, 41]. A comparison of the pollen sterility frequency distribution is presented in **Figure 5** for 2008S and 2009S F_2 populations.

3.2 Candidate genes and mechanism of genetic male sterile rice

Several candidate genes were identified in controlling male sterility in PGMS lines. More recent studies using Nongken 58S discovered that the gene in LOC_Os12g36030 which was previously mapped as pms3 and is also allelic to p/tms12-1 located on chromosome 12 regulates photoperiod genetic male sterility. The single nucleotide polymorphism (SNP) mutation located in non-coding region increased the methylation on the promoter which in turn reduced transcription of LOC_Os12g36030.The reduced transcription caused the pre-programmed cell death in developing anthers causing pollen sterility [42, 43].

A second study conducted in PTGMS line Pei'ai 64S revealed that the gene LOC_Os07g12130 previously mapped as pms1(t) encodes a protein containing Myb-like DNA binding domain that affect the transcription of a protein responsible for the photo-thermosensitive response. RT-PCR results showed that mRNA levels of LOC_Os07g12130 changes at different photoperiod and temperature conditions [31]. However, the gene is yet to be cloned and further study needs to be conducted. De Guzman [4] sequenced both locus in male sterile line 2008S and found both SNPs present. When QTLs were analyzed in two segregating populations using both single marker analysis and interval mapping, each locus and their interaction gives significant effects [25].

For lines exhibiting thermosensitivity, candidate genes associated with TGMS lines were mapped on chromosome 2. QTL mapping using bulk segregant analysis approach (BSA) identified the ptgms2-1 locus [30]. Further analysis showed

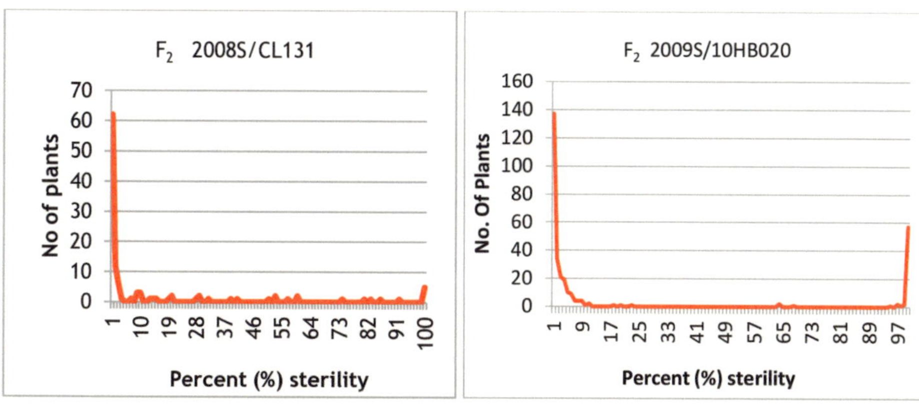

Figure 5.
Pollen sterility distribution of F2 plants in populations of 2008S/CL131 and 2009S/10HB020 male sterile lines planted in 2013 at H. Rouse Caffey Rice Research Station, Crowley LA USA.

LOC_Os02g12290 encode a ribonuclease Z gene that when the SNP is present, it created a premature stop codon rendering the RNase Z^{S1} defective. The mechanism described that when the mutant male sterile is exposed under high temperature (28°C), it induced the accumulation of mRNA ribosomal protein UbL40 in microspore mother cell that were not processed by the defective RNase Z^{S1} enzyme consequently causing pollen degeneration. At lower temperatures (23°C), UBL40 mRNAs levels remained low allowing production of normal pollen. This mutation was reported in TGMS varieties Guangzhang 63S, Ans-1 and Zhu1S [28]. Zhang [44] discovered that the locus tms5, ptgms2-1 and tms9 were allelic and were all mapped to chromosome 2 that contains the similar ribonuclease Z gene. De Guzman [25] sequenced the locus LOC_Os02g12290 in line 2009S and discovered the same SNP present in TGMS lines. Inheritance studies on F_2 and BC_1F_2 showed similar segregation ratios. SNP marker were developed using CEL1 nuclease to identify association of the marker to the trait and showed that the markers were able to predict 95–100% of male sterile lines in F_2 and BC_1F_2 population [41].

3.3 Breeding and production of genetic male sterile rice

Since PTGMS and TGMS have different responses to temperature and photoperiod, methods in seed multiplication and breeding varies including selection of specific location and time of planting. On PTGMS lines, weak and strong photoperiodic responses were reported [45]. In strong photoperiodic response such as in long daylength, the critical sterility inducing temperature (CST) is low (21°C) and at short daylength high (25–26°C).

For a PTGMS lines with weak photoperiodism, the (CST) is about 22°C under short daylength. In China, Chen [45] suggested that seed production of this type should be bred during autumn in Guangdong and Guangxi, and in winter in Hainan province. Cold water irrigation treatment has been used extensively in lines with both weak and strong photoperiod. This solved the problem of low yields in multiplication of PTGMS lines such as Pei'-ai 64 s with low CST [46].

For TGMS lines, there is no weak or strong photoperiodism, thus timing and selection of location is important for seed production of male steriles and hybrids. On most TGMS lines, the ideal CST is about 22.5°C. Seed production are also treated with continuous cold water irrigation to increase seed yield during the winter [46].

In the US, RiceTec of Alvin Texas successfully used both ptgms and tgms lines however, seed production locations are unknown and specific methods of seed multiplications of male sterile lines are undisclosed.

In Louisiana USA, the LSU AgCenter initiated the hybrid rice breeding program in 2009 using ptgms and cytoplasmic male sterile lines obtained via a Material Transfer Agreement with the Guangxi Academy of Agricultural Sciences, Nanning, China. Test crosses made from these lines showed equal or superior grain and head rice yield compared to the current RiceTec commercial varieties. However, high chalk, lodging, and late maturities were observed that warrant the development of male sterile with improved agronomic traits and are suitable to the southern US conditions [47]. The breeding cycle starts with crosses of adapted lines to male sterile lines to produce F_1's. Seeds from hybrids then harvested and spaced planted. Single plant selections of male sterile lines were performed by looking at sterile pollen stained with 1% I_2KI under the microscope during early heading. Plants suitable for generation advance were selected with the following characteristics: 98–100% pollen sterile, 60–80 cm in height, short flag leaf, intermediate tillering and with compact growth habit. Selected plants were uprooted, placed in one gallon pots and transferred in the greenhouse. The plants were ratooned by cutting ~9–10 cm from the soil line. The ratoons are allowed to grow up to early booting stage where the

Figure 6.
Methods of selection and production of male sterile lines in LSU Agcenter hybrid rice breeding program.

majority of the shoots have the measurement distance of about of 1–3 cm between the flag leaf collar to the leaf node or at about the early booting stage (meiotic division of pollen mother cell). Plants are then subsequently treated in a growth chamber with temperatures 22°C during the night and 28°C during the day at 11 h daylength for 10 d. After treatment, plants are then transferred to the greenhouse for 30–35 days for the seed to mature (**Figure 6**). Seed multiplication were carried out in Puerto Rico agricultural experiment station planted during October to November [48, 49].

4. Introgression of disease resistance in hybrid rice

Introgression of disease resistance traits in hybrid rice becomes a necessity largely due to the narrow genetic diversity of both CMS and genetic male sterile sources. A study showed leading hybrids from China that was introduced in Africa were out yielded by inbred checks due to non-adaptability and susceptibility to diseases and pests [50]. There were also reported insect and disease incidence in China that are more frequent in hybrid rice than on inbred varieties [5]. Research institutes such as the International Rice Research Institute (IRRI) was aware of these issues and has continued to develop new CMS, maintainer, restorers and genetic male steriles in diverse background. Current improvement on hybrid rice focuses on incorporation of resistance gene identified from inbred and wild sources. For instance, blast resistance genes have been introgressed in maintainer, restorer and S lines through hybridization, backcross and marker assisted selection (MAS) [6]. In the early 2000, varieties with multiple bacterial panicle blight resistance were released in Indonesia and China. These varieties were developed using MAS and produced significant yield gain demonstrated in farmer's field [51].

5. Future aspect

CMS and genetic male sterility revolutionize rice production due to its contribution to the development of hybrid rice. Elucidation of physiological and molecular mechanism leads to the establishment of the process involved in breeding male sterile and hybrids. The increase in yield and tolerance to biotic and abiotic stress are largely due to the effect of heterosis. However, the narrow genetic diversity presents a challenge as very few sources of male sterility are used. Improving parental lines by incorporation of genes with resistance to biotic and abiotic conditions are essential to secure the yield advantage of hybrids over inbreds. Different methods can be used to add genetic diversity to the hybrids. Mutation, MAS and interspecific hybridization are a proven approach to incorporate targeted traits as well as discovering new genes from wild relatives. Genomic selection as well as gene editing will likely play a significant role in future improvement of rice hybrids.

Acknowledgements

This work was supported by Southeast Missouri State University and the Missouri Rice Research and Merchandising Council. The authors also wish to thank the assistance of the LSU Agcenter H. Rouse Caffey Rice Research Station for research related to two line hybrids.

Conflict of interest

The authors declare that they have no conflict of interests.

Author details

Christian De Guzman[1*] and James Oard[2]

1 Southeast Missouri State University, Cape Girardeau, Missouri, USA

2 LSU Agcenter, H. Rouse Caffey Rice Research Station, Crowley, Louisiana, USA

*Address all correspondence to: cdeguzman@semo.edu

References

[1] Yuan LP. Current Status of the Two-line Hybrid Rice Research. China: Agricultural Press; 1992. p. 377

[2] Zhang Q et al. Using bulked extremes and recessive class to map genes for photoperiod-sensitive genic male sterility in rice. Proceedings of the National Academy of Sciences of the United States of America. 1994;**91**(18):8675-8679

[3] Chen L, Liu YG. Male sterility and fertility restoration in crops. Annual Review of Plant Biology. 2014;**65**:579-606

[4] De Guzman C et al. Genetic analysis of photoperiod/thermosensitive male sterility in Rice under US environments. Crop Science. 2017;**57**(4):1957-1965

[5] Virmani SS. Disease and insect resistance in hybrid rice. In: Virmani SS, editor. Heterosis and Hybrid Rice Breeding. Berlin, Heidelberg: Springer; 1994. pp. 111-118

[6] Xing J, Deng H, Yuan L. The utilization of rice blast resistance genes in hybrid rice breeding in China. IntechOpen. 24 January 2019. Available from: https://www.intechopen.com/online-first/the-utilization-of-rice-blast-resistance-genes-in-hybrid-rice-breeding-in-china

[7] Wang LJSB. Yuan Longping: Hybrid rice is on the way to fulfilling its potential. Science Bulletin. 2015;**60**(6):657-660

[8] Cheng S-H et al. Progress in research and development on hybrid rice: A super-domesticate in China. Annals of Botany. 2007;**100**(5):959-966

[9] Huang JZ et al. Workable male sterility systems for hybrid rice: Genetics, biochemistry, molecular biology, and utilization. Rice (New York, N.Y.). 2014;**7**(1):13

[10] Shen X-h et al. Development of novel cytoplasmic male sterile source from Dongxiang wild rice (*Oryza rufipogon*). Rice Science. 2013;**20**:379-382

[11] Li S, Wan C, Zhu Y. Characterization and use of male sterility in hybrid rice breeding. Journal of Integrative Plant Biology. 2007;**49**:791-804

[12] Fujii S, Toriyama K. Suppressed expression of retrograde-regulated male sterility restores pollen fertility in cytoplasmic male sterile rice plants. Proceedings of the National Academy of Sciences of the United States of America. 2009;**106**(23):9513-9518

[13] Li S, Yang D, Zhu Y. Characterization and use of male sterility in hybrid rice breeding. Journal of Integrative Plant Biology. 2007;**49**:791-804

[14] Wang Z. Et al, Cytoplasmic male sterility of rice with Boro II cytoplasm is caused by a cytotoxic peptide and is restored by two related PPR motif genes via distinct modes of mRNA silencing. Plant Cell. 2006;**18**(3):676-687

[15] Virmani SS et al. Two-Line Hybrid Rice Breeding Manual. Los Baños (Philippines): IRRI; 2003. p. 88

[16] Akagi H et al. A unique sequence located downstream from the rice mitochondrial atp6 may cause male sterility. Current Genetics. 1994;**25**(1):52-58

[17] Luo D et al. A detrimental mitochondrial-nuclear interaction causes cytoplasmic male sterility in rice. Nature Genetics. 2013;**45**:573

[18] Delannoy E et al. Pentatricopeptide repeat (PPR) proteins as sequence-specificity factors in post-transcriptional processes in organelles. Biochemical Society Transactions. 2007;**35**(6):1643-1647

[19] Huang W et al. Two non-allelic nuclear genes restore fertility in a gametophytic pattern and enhance abiotic stress tolerance in the hybrid rice plant. Theoretical and Applied Genetics. 2011;**124**:799-807

[20] Tang H et al. The Rice restorer Rf4 for wild-abortive cytoplasmic male sterility encodes a mitochondrial-localized PPR protein that functions in reduction of WA352 transcripts. Molecular Plant. 2014;**7**(9):1497-1500

[21] Ariful I et al. Development of component lines (CMS, maintainer and restorer lines) and their maintenance using diversed cytosources of rice. Rice Research. 2015;**3**(3):5

[22] Huang F et al. Genetically characterizing a new *indica* cytoplasmic male sterility with cytoplasm for its potential use in hybrid rice production. Crop Science. 2013;**53**:132

[23] Yuan LP. Progress of two-line system hybrid rice breeding. Scientia Agricultura Sinica. 1990;**23**(3):6

[24] Cheng SH et al. Classification of environmentally induced genetic male sterile lines of rice based on their fertility responses to photoperiod and temperature. The Journal of Agricultural Science. 1996;**127**(Issue 02):161-167

[25] De Guzman C. Genetic analyses of male sterility and wide compatibility in U.S. hybrid rice breeding lines [doctoral dissertation]. Baton Rouge Louisiana: Louisiana State University; 2016

[26] Mei G, Wang X, Wang M. Genetic analysis of the photoperiod sensitive male sterility of Nongken 58S and its derivatives. Journal Huangzhong Agricultural University. 1990;**9**(4):400-406

[27] Chen LB, Zhou GQ, Huang YX. Influence of temperature and photoperiod on the fertility and physiology of Annong S-1 and Hengnong S-1 of rice. Acta Botanica Sinica. 1994;**36**(Suppl):119-123

[28] Zhou H et al. RNase Z(S1) processes UbL40 mRNAs and controls thermosensitive genic male sterility in rice. Nature Communications. 2014;**5**:4884

[29] Qi Y et al. Fine mapping and candidate gene analysis of the novel thermo-sensitive genic male sterility tms9-1 gene in rice. Theoretical and Applied Genetics. 2014;**127**(5):1173-1182

[30] Xu J et al. Fine mapping and candidate gene analysis of ptgms2-1, the photoperiod-thermo-sensitive genic male sterile gene in rice (*Oryza sativa L.*). Theoretical and Applied Genetics. 2011;**122**(2):365-372

[31] Zhou YF, Zhang XY, Xue QZ. Fine mapping and candidate gene prediction of the photoperiod and thermo-sensitive genic male sterile gene pms1(t) in rice. Journal of Zhejiang University. Science. B. 2011;**12**(6):436-447

[32] He YQ et al. Genetic bases of instability of male sterility and fertility reversibility in photoperiod-sensitive genic male-sterile rice. Theoretical and Applied Genetics. 1999;**99**(3-4):683-693

[33] Yang RC et al. A recessive gene in *indica* rice 5460S for thermosensitive genic male sterility. Rice Genetics Newsletter. 1992;**9**:56-57

[34] Mei MH et al. pms3 is the locus causing the original photoperiodsensitive male sterility mutation of "Nongken 58S". Science in China. (Series C.). 1999;**42**(3):316-322

[35] Mei MH et al. Mapping and genetic analysis of the genes for photoperiod-sensitive genic male sterility in rice using the original mutant nongken 58S. Crop Science. 1999;**39**(6):1711-1715

[36] Lopez MT et al. Microsatellite markers flanking the tms2 gene facilitated tropical TGMS rice line development. Crop Science. 2003;**43**:2267-2271

[37] Peng HF et al. Fine mapping of a gene for non-pollen type thermosensitive genic male sterility in rice (*Oryza sativa L.*). Theoretical and Applied Genetics. 2010;**120**(5):1013-1020

[38] Dong VN et al. Molecular mapping of a rice gene conditioning thermosensitive genic male sterility using AFLP, RFLP and SSR techniques. Theoretical and Applied Genetics. 2000;**100**(5):727-734

[39] Subudhi PK et al. Molecular mapping of a thermo-sensitive genetic male sterility gene in rice using bulked segregant analysis. Genome. 1997;**40**:188-194

[40] Alcochete et al. Mapping of quantitative trait loci for thermosensitive genic male sterility in *indica* rice. Pesquisa Agropecuária Brasileira. 2005;**40**(12):1179-1188

[41] De Guzman CT, Linscombe SD, Oard J. Genetic analysis of environment sensitive genic male sterile rice under US environments. Euphytica. 2019;**215**(2):3941

[42] Ding J et al. A long noncoding RNA regulates photoperiod-sensitive male sterility, an essential component of hybrid rice. Proceedings of the National Academy of Sciences of the United States of America. 2012;**109**(7):2654-2659

[43] Ding J et al. RNA-directed DNA methylation is involved in regulating photoperiod-sensitive male sterility in rice. Molecular Plant. 2012;**5**(6):1210-1216

[44] Zhang H-L et al. Characterization of an RNase Z nonsense mutation identified exclusively in environment-conditioned genic male sterile rice. Molecular Breeding. 2014;**34**(2):481-489

[45] Chen LL, Tang DW, Xiao Y. Mechanism of sterility and breeding strategies for ptgms. Rice Science. 2010;**17**(3):161-167

[46] Li J, Xin Y, Yuan L. Hybrid rice technology development: Ensuring China's Food Security. IFPRI Discussion Paper 00918. 2009. p. 1-28

[47] Casella FM. Development and evaluation of environmentally-sensitive genetic male sterile lines for Louisiana hybrid rice [doctoral dissertation]. Baton Rouge Louisiana: Louisiana State University; 2017

[48] De Guzman C, Oard J, Galam D, Camacho R, Sanabria Y, Esguerra M, et al. Predicting pollen sterility using SNP markers in single and two gene recessive inheritance of photoperiod and thermosensitive genetic male sterile rice. In: ASA/CSSA/SSSA Annual Meeting. 2015. 702 p

[49] Casella FM. Development and evaluation of environmentally-sensitive genetic male sterile lines for Louisiana hybrid rice [doctoral dissertation]. Baton Rouge Louisiana: Louisiana State University; 2017

[50] Virmani SS, Kumar I. Development and use of hybrid rice technology to increase rice productivity in the tropics. International Rice Research Institute. 2004;**29**:10-20

[51] Leung H et al. Sustainable disease resistance in rice: Current and future strategies. New directions for a diverse planet: In Proceedings of the 4th International Crop Science Congress. Brisbane, Australia. 26 Sep-1 Oct 2004. http://www.cropscience.org.au/icsc2004/pdf/1022_leungh.pdf

Chapter 3

The Utilization of Rice Blast Resistance Genes in Hybrid Rice Breeding in China

Junjie Xing, Huafeng Deng and Longping Yuan

Abstract

Hybrid rice has demonstrated promises of yield gain for over several decades since its conception and massive deployment in China. One of the common bottlenecks of hybrid rice is the availability of suitable breeding lines as parents to produce marketable rice grains. Due to limitation of genetic diversity of breeding parent, hybrid rice is extremely vulnerable to rice blast disease caused by the fungal pathogen *Magnaporthe oryzae*. *M. oryzae* is a highly adaptive fungus that often gains new virulence to reduce crop resistance resulting in massive yield loss and crop failure. To secure yield gain of hybrid rice, identification and integration of diverse sources of resistance genes into hybrid rice are super critical. In this chapter, we will present strategies to identify, characterize, and stack effective blast resistance genes in hybrid rice breeding in China.

Keywords: rice blast, resistance gene, hybrid rice

1. Introduction

In China, the research on hybrid rice has gone on for more than 50 years. Professor Yuan first found the male sterility in 1964 and started hybrid rice research in China and, subsequently, creatively proposed the three-line, two-line, and one-line breeding conception [1]. Three-line hybrid rice was defined as restorer line, cytoplasmic male sterile line, and maintainer line; two-line hybrid rice was defined as restorer line and photo-thermosensitive genic male sterile; one-line hybrid rice was defined to maintain the heterosis by diploid line through apomixes [2]. Until now, hybrid rice breed with three-line or two-line method has successfully been applied in rice production.

From 1975, hybrid rice has gone through fast-speed development. More than 5000 varieties have been authorized by the government and planted for more than 500 million hm^2 in China and play important function for national food safety [3, 4]. Rice blast disease caused by *Magnaporthe oryzae* is popular and devastative on rice. The vulnerability of hybrid rice to rice blast brought huge yield damage. The utilization of resistant varieties was the most economical and environmental method to control the rice blast. Up to now, more than 90 resistance (*R*) genes have been identified, in which more than 20 genes are cloned [5]. Hence, rice lines containing major resistant genes have been widely used directly or indirectly as parents of hybrid rice. In this chapter, we will introduce the utilization of resistance genes in hybrid rice breeding.

2. The utilization of rice blast resistance genes

2.1 Identification of the resistant rice parents

The resistance level of parents is directly related with the resistance performance of hybrid rice. The resistance evaluation for breeding lines is a very important prerequisite work for resistance breeding of hybrid rice. For traditional breeders, field nursery or artificial inoculation with blast isolates in greenhouse was normally used for resistance identification. Amounts of rice lines with middle or high resistance have been identified in different provinces with diverse ecology. The detailed information was listed in **Table 1**. These identified rice materials provided rich selection as parents or resistant resource for hybrid rice breeding. As we know, genetic mechanism of rice blast resistance followed the gene for gene interaction. It was unclear about background and resistant genes in these materials, and the presence of one *R* gene masked another *R* gene; and also, the stationary field nursery only can stand for limited ecological districts. Hence, blast evaluation cannot identify any particular resistance gene, and it will lead to huge uncertainty in resistance of later generations in breeding. Phenotype identification cooperated with precise analysis of resistant genes will more effectively serve for hybrid rice resistance breeding.

2.2 Characterization and stack effective *R* genes in hybrid rice

Following the clone of resistant genes and the development of related functional molecular makers, marker screening has been widely applied on resistance identification and innovation of parents of hybrid rice. Up to now, molecular makers of blast resistance genes, *Pi2*, *Pi9*, *Pi1*, *Pib*, *Pita*, *Pid2*, *Pikh*, *Pigm*, and *Pish*, have been developed and used on detecting the related *R* genes [17–25]. In Jiangsu, a total of 544 rice materials were assessed for blast resistance and resistance genes distribution by inoculation and marker screening; results showed that 968, Xiushui 134, Jia 58, Jindao 263, Huaidao 20, Yandao 10, and Gumei 4 contained the majority of resistance genes; and *Pi5*, *Pita*, *Pi9*, and *Pib* exhibited high resistance to six major blast races [14]. In South China, with functional marker of *Pi1*, *Pik-p*, *Pikh*, *Pi2*, *Pi9*,

Province	Resistance resource	Reference
Sichuan	IR99–35, Miyang 46, IR 1544, Tetep, Gumei 2, 6326, Suhui162, and Suhui 527	[6, 7]
Heilongjiang	Suijing 12, Mudanjiang 26, Longdun 105, Longjing 20, Longjing 31, Dongnong 415, Songjing 9, Longdao 12, Hejiang 23, and Wuyoudao 3	[8]
Guangdong	Sanhuangzhan 2 Hao, Qingliuai 1 Hao, Jingxian 89, IR36, and 28 Zhan	[9]
Hunan	Xiangzao 143, Fengyuanyou 299, Jinyou 207, Liangyou 222, Quanfengyou 610, Hanyou 983, Lvyinzhan, Bingyou C278, Yuenongsimiao, and Zhuoliangyou 249	[10, 11]
Hubei	Zhenke, Jinlong 1, Fanyu 1, Ningwan 1, Sanqizao, Nanjing 15, Aiyinnuo, Jinzao 47, Yunjin 23, and Quanzhen 10	[12, 13]
Fujian	Yixiangyou 673, Dyou 15, Gangyou 148, Guyou 527, Jiafuzan, and Teyou 627	[14, 15]
Jiangsu	Longjing 968, Xiushui 134, Jia 58, Jindao 263, Huaidao 20, Yandao 10, and Gumei 4	[16]

Table 1.
Selected rice varieties with different resistant resources to blast in China.

Piz-t, *Pita*, and *Pii*, 328 hybrid rice combinations were screened, in which *Pita* and *Pii* were found in high frequency, but *Pi2* and *Pi1* displayed highly effective resistance contribution to local rice [26]. In Sichuan Province, with molecular markers closely linked to *Pi-9*, *Pi-2*, *Pi-kh*, and *Pi-km*, general rice parents of hybrid rice were analyzed and selected for the resistance resources [27]. The *R* gene screening make breeders directly utilize related resistance resources on purpose.

The hybridization, backcross, and marker-assisted selection (MAS) were the general method for the introduction of *R* genes into the restorer line, maintainer line, and sterile line of hybrid rice. MAS conduced to selectively breeding based on the genotype and accelerate the breeding course [28]. The procedure for MAS was shown in **Figure 1**. As we know, *R* genes, such as *Pi9*, *Pi2*, *Pi1*, and *Pigm*, have been reported to show relatively high resistance in different districts in China [29–31]. Hence, these *R* genes were often used for improvement of rice blast

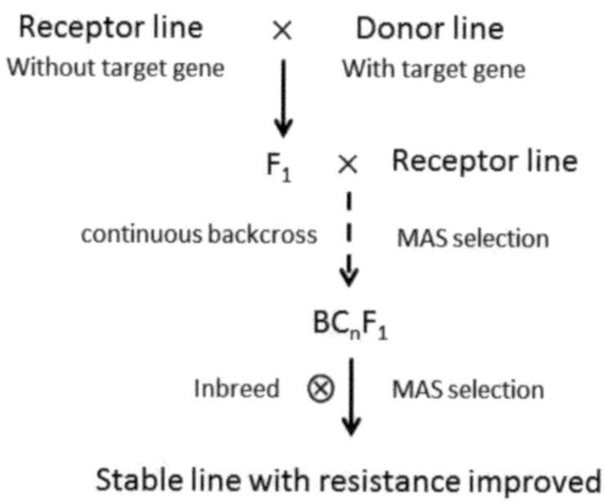

Figure 1.
The breeding course with marker-assisted selection and backcross.

R genes used	Variety improved	Variety type	Reference
Pi9	Yandao 6 Hao	General cultivar	[32]
Pi25	Xiangwanxian 13	General cultivar	[33]
Pi1, *Pi2* and *Pi33*	Jin 23B	Maintainer line	[34]
Pi1 and *Pi2*	Rongfeng B	Maintainer line	[35]
Pid(t), *Pib* and *Pita*	G46B	Maintainer line	[36]
Pi9	R599	Maintainer line	[37]
Pi9	R288	Maintainer line	[38]
Pi9	Shuhui527, Minghui 86, and Minhui 3301	Restorer line	[39]
Pigm(t)	Chunhui 350	Restorer line	[40]
Pi9 and *Pi49*	Chuang 5S	Sterile line	[41]
Pi25	Zhenda A	Sterile line	[42]
Pi47 and *Pi48*	C815S	Sterile line	[43]
Pi9	03S	Sterile line	[44]
Pi1	GD-8S	Sterile line	[45]
Pi1 and *Pi2*	Peiai64S	Sterile line	[46]

Table 2.
The improved rice varieties of different types with MAS technique.

resistance (**Table 2**). Recently, a new class resistance gene *Ptr* just cloned encoded an atypical protein and conferred broad-spectrum disease resistance and will provide diverse selection for resistance improvement [47]. To breeding rice cultivars with durable blast resistance, stacking several resistance genes still was the most effective method. To stack resistance gene purposefully, spectrum of each resistance gene must be determined, and also, the identification of differential blast races/isolates that distinguished each resistance gene in different districts was critical for ensuring the effectiveness of resistance gene stacking [48].

3. Conclusions

With identification in the rice blast field nursery or functional marker detecting of major *R* genes, the amount of blast resistance resources was identified and provides diverse selections for hybrid rice resistance breeding. However, the recent finding showed that *Pita* required *Ptr* to function revealed that part of single *R* gene may be not functional as we thought originally [47], and further function analysis of more *R* genes may be necessary.

For conventional rice breeding, all blast *R* genes must be stacked into breeding lines to be effective, whereas hybrid rice can stack blast *R* genes into two parents. For hybrid rice breeding, blast resistance was only a part of index, and other agronomic traits also need to be considered. Lines of Chuang 5 S stacked *Pi9* and *Pi49* have been found obvious differences on plant height, spike length, spike number and stigma exertion rate with the receptor, even though the blast resistance has improved [41]. Hence, blast resistance breeding for hybrid rice was a synthetic work that contained resistance innovation and excellent agronomic trait selection.

In this chapter, it introduced the progress on identification of resistance resources and the utilization of blast resistance genes. Traditional cross technique, combined with MAS, has been used to transfer different major *R* genes into parent's lines to improve the resistance of hybrid rice and achieved remarkable results. Following the improvement of blast resistance of the authorized varieties, it gratefully contributed to decrease the damage of rice blast disease and played important function on protection of rice production safety in China.

Acknowledgements

This work was supported by Hunan Provincial Science and Technology Department (CN) (2018NK1020), Hunan Academy of Agricultural Science and Technology Innovation Project (2018ZD01-7), Hunan Academy of Agricultural Science and Technology Innovation Project (2017JC03), and Hunan Provincial Natural Science Fund Youth Fund (2018JJ3379).

Conflict of interest

The authors declared that there was no conflict of interest.

Author details

Junjie Xing*, Huafeng Deng and Longping Yuan
Hunan Hybrid Rice Research Center, Changsha, Hunan, China

*Address all correspondence to: xingjunjie@hhrrc.ac.cn

References

[1] Wu B, Hu W, Xing YZ. The history and prospect of rice genetic breeding in China. Hereditas. 2018;**40**:841-857

[2] Yuan LP. The strategy for hybrid rice development. Hybrid Rice. 2018;**33**:1-2

[3] Zeng B, Sun SX, Wang J. Registration of main rice varieties and its application in recent 30 years in China. Crops. 2018;**2**:1-5

[4] Hu ZX, Tian Y, Xu QS. Review of extension and analysis on current status of hybrid rice in China. Hybrid Rice. 2016;**31**:1-8

[5] Yi NA, Li W, Dai LY. Advances in the cloning of rice blast resistance gene and its molecular breeding. Molecular Plant Breeding. 2015;**13**:1653-1659

[6] Huang F, Ye HZ, Xie R, Liu CY. Screening for good quality rice germplasm resources resistant to rice blast. Acta Agronomica Sinica. 2006;**32**:1549-1553

[7] Ma BT, Yang L, Wang LX, Yang TZ, Zhao XY, Gao XG, et al. Resistance evaluation and breeding of rice varieties (lines) to rice blast in nursery. Chinese Agricultural Science Bulletin. 2005;**21**:263-267

[8] Xin W, Wang JG, Sun J, Liu HL, Guo LY, Jiang SD, et al. Physiological races of rice blast in Heilongjiang province and species identification of resistance to resource. Acta Agriculturae Boreali-Sinica. 2016;**31**:130-137

[9] Feng AQ, Yang JY, Chen S, Zeng LX, Yang QY, Su J, et al. Evaluation on qualitative and quantitative resistance of rice germplasm to *Magnaporthe oryzae*. Guangdong Agricultural Sciences. 2015;**42**:27-32

[10] Feng GP, Liu EM, Huang HM, Lv JL, Yan WT, Bai ZA, et al. Identification of resistance of 121 rice varieties against *Magnaporthe grisea*. Hybrid Rice. 2010;**25**:79-81

[11] Min J, Mao SC, Long HP, Yi GL, Sun MY. Results and analysis of blast resistance of rice varieties in the united comparison test of rice in Hunan province. Hybrid Rice. 2018. DOI: 10.16267/j.cnki.1005-3956.20180322.091

[12] Zeng FS, Xiang LB, Yang LJ, Yang XL, Yang JS, Yu DZ. Diversity analysis for resistance of 251 rice (*Oryza sativa* L.) varieties (lines) to rice blast disease. Acta Phytopathologica Sinica. 2011;**41**:399-410

[13] Li XS, Xiang XJ, Shen CC, Yang LW, Chen K, Wang XW, et al. Identification and evaluation of blast resistance for resequenced rice core collections. Acta Agronomica Sinica. 2017;**43**:795-810

[14] Ruan HC, Yang XJ, Chen SL, Dong RX, Chen FR, Wang WX, et al. Identification and evaluation of rice new varieties resistant to *Magnaporthe grisea* in Fujian Province. Fujian Journal of Agricultural Sciences. 2006;**21**:304-307

[15] Du YX, Ruan HC, Wang MM, Guan RF, Yang XJ, Gan L, et al. Resistance evaluation of leading rice varieties to rice blast disease in Fujian province. Chinese Agricultural Science Bulletin. 2010;**26**:217-221

[16] Li G, Yuan CY, Cao KR, Sun XL, Li J, Wang J, et al. Evaluation and distribution of the blast resistance genes of 544 rice materials. Journal of China Agricultural University. 2018;**23**:22-28

[17] Jiang JS, Wang SP. Identification of a 118kb DNA fragment containing the locus of blast resistance gene Pi2(t) in rice. Molecular Genetics and Genomics. 2008;**268**:249-252

[18] Qu SH, Liu GF, Zhou B, Bellizzi M, Zeng LR, Dai LY, et al. The broad-spectrum blast resistance gene Pi9 encodes a nucleotide-binding site leucine-rich-repeat protein and a member of a multi gene family in rice. Genetics. 2006;**172**:1901-1904

[19] Hua LX, Wu JZ, Chen CX, Wu WH, He XY, Lin F, et al. The isolation of Pi1, an allele at the Pik locus which confers broad spectrum resistance to rice blast. Theoretical and Applied Genetics. 2012;**125**:1047-1055

[20] Liu Y, Xu PZ, Zhang HY, Xu JD, Wu FQ, Wu XJ. Marker-assisted selection and application of blast resistant gene Pib in Rice. Scientia Agricultura Sinica. 2008;**41**:9-14

[21] Jia YL, Wang ZH, Singh P. Development of dominant rice blast Pi-ta resistance gene markers. Crop Science. 2002;**42**:2145-2149

[22] Gao LJ, Deng GF, Gao HL, Gao GQ, Zhou M, Zhou WY, et al. Establishment and application of gene tagging linked to rice blast resistance Gene Pi-d2. Southwest China Journal of Agricultural Sciences. 2010;**23**:77-82

[23] Ramkumar G, Srinivasarao K, Madhan Mohan K, Sudarshan I, AKP S, Gopalakrishna K, et al. Development and validation of functional marker targeting an InDel in the major rice blast disease resistance gene Pi54 (Pikh). Molecular Breeding. 2011;**27**:129-135

[24] Zeng SY, Li C, Du CC, Sun LT, Jing DD, Lin TZ, et al. Development of specific markers for Pigm in marker-assisted breeding of panicle blast resistant Japonica rice. Chinese Journal of Rice Science. 2018;**32**:453-461

[25] Akira T, Nagao H, Akio M, Hirohiko H. Unique features of the rice blast resistance Pish locus revealed by large scale retrotransposon tagging. BMC Plant Biology. 2010;**10**:175

[26] Wang WJ, Zhou JY, Wang CY, Su J, Feng JQ, Chen B, et al. Distribution of eight rice blast resistance genes in Indica hybrid rice in China. Chinese Journal of Rice Science. 2017;**31**:299-306

[27] Feng H, Yang CM, Wu XB, Liu YS, Peng YL. Detection and analysis of rice blast resistance gene in some hybrid rice parents and 32 resistant materials of Sichuan province. Southwest China Journal of Agricultural Sciences. 2013;**26**:987-993

[28] Wang J, Yang J, Chen ZD, Fan FJ, Zhu JY, Yang JH, et al. Pyramiding resistance gene Pi-ta, Pi-b, and Stv-bi by marker-assisted selection in rice (*Oryza sativa* L.). The Crop Journal. 2011;**37**:975-981

[29] Chen HL, Chen BT, Zhang DP, Xie YF, Zhang Q. Pathotypes of *Pyricularia grisea* in rice fields of central and southern China. Plant Disease. 2001;**85**:843-850

[30] Liu G, Lu G, Zeng L, Wang GL. Two broad-spectrum blast resistance genes, Pi9 (t) and Pi2 (t), are physically linked on rice chromosome 6. Molecular Genetics and Genomics. 2002;**267**:472-480

[31] Deng YW, Zhai K, Xie Z, Yang DY, Zhu XD, Liu JZ, et al. Epigenetic regulation of antagonistic receptors confers rice blast resistance with yield balance. Science. 2017;**355**:962

[32] Yin DS, Xia MY, Li JB, Wan BL, Zha ZP, Du XS, et al. Development of STS marker linked to rice blast resistance gene Pi9 in marker-assisted selection breeding. Chinese Journal of Rice Science. 2011;**25**:25-30

[33] Liu WQ, Li XX, Li YC, Pan XW, Sheng XN, Duan YH. Improvement of rice blast resistance of Xiangwanxian No.13 with high quality by molecular marker-assisted selection. Molecular Plant Breeding. 2017;**15**:3063-3069

[34] Chen HQ, Chen ZX, Ni S, Zuo SM, Pan XB, Zhu XD. Pyramiding three genes with resistance to blast by marker-assisted selection to improve rice blast resistance of Jin 23B. Chinese Journal of Rice Science. 2008;**22**:23-27

[35] Liu WG, Wang F, Liu ZR, Zhu XY, Li JH, Huang HJ, et al. Improvement of rice blast resistance in CMS line Rongfeng A by pyramiding Pi-1 and Pi-2 with molecular marker techniques. Molecular Plant Breeding. 2012;**10**:575-582

[36] Chen XW, Li SG, Ma YQ, Li HY, Zhou KD, Zhu LH. Marker-assisted selection and pyramiding for three blast resistance genes, Pi-d(t)1, Pi-b, Pi-ta2 in Rice. Chinese Journal of Biotechnology. 2004;**20**:708-714

[37] Chen HL, Li YC, Liu XL, Liu JL, Yang FY, Xing X, et al. Improving blast resistance of Indica rice restorer R599 and its hybrid by MAS technology. Hybrid Rice. 2017;**32**:61-65

[38] Xing X, Liu XL, Chen HL, Yang FY, Li YC, Liao H, et al. Improving blast resistance of rice restorer R288 by molecular marker-assisted selection of Pi9 Gene. Crop Research. 2016;**30**:487-491

[39] Zhang H, Zhou P, Tu SH, Zheng JT, Zhang JF, Xie HA. Developing new restorer lines with blast-resistance gene Pi9 for hybrid rice by marker assistance selection (MAS). Molecular Plant Breeding. 2015;**13**:1918-1922

[40] Yang P, Zou GX, Chen CL, Huang YP, Lan B, Xiong YH, et al. Improvement of rice blast resistance of Chunhui350 by using molecular-marker assisted selection. Molecular Plant Breeding. 2015;**13**:741-747

[41] Zhang JP, Hao M, Zeng G, Cao Z, Jiang HY, Huang XG, et al. Polymerization of Pi9 and Pi49 loci by marker assisted selection to improve blast resistance of dual-purpose genic sterile rice Chuang 5S. Molecular Plant Breeding. 2018;**16**:7372-7379

[42] Dong RX, Wang HF, Dong LF, Zhou P, Tu SH, You QR, et al. Improving the rice blast resistance for a CMS line of rice Zhenda A and its hybrids using molecular marker-assistant selection. Journal of Plant Genetic Resources. 2017;**18**:573-586

[43] Cao Z, Zeng G, Hao M, Sheng HW, Ye NZ, Xiao YH. Improving blast resistance of dual-purpose genic sterile line C815S by using molecular marker-assisted selection. Molecular Plant Breeding. 2015;**13**:1193-1200

[44] Wang H, Chen JJ, Yan Z, Zhou GX, Shen GL, Jiang JP, et al. Improving the rice blast resistance of sterile line 03S by molecular marker-assisted selection. Journal of Yangzhou University. 2015;**36**:74-78

[45] Jin SJ, Liu WG, Zhu XY, Wang F, Li JH, Liu ZR, et al. Improving blast resistance of a thermo-sensitive genic male sterile line GD-8S by molecular marker-assisted selection. Chinese Journal of Rice Science. 2007;**21**:599-604

[46] Dong W, Li X, Yan B, Wu CJ, Gao GJ, Bao L, et al. Improving the blast resistance of Peiai64S through marker-assisted selection. Molecular Plant Breeding. 2010;**8**:853-860

[47] Zhao HJ, Wang XY, Jia YL, Minkenberg B, Wheatley M, Fan JB, et al. The rice blast resistance gene Ptr encodes an atypical protein and confers broad spectrum disease resistance. Nature Communications. 2018;**9**:2039. DOI: 10.1038/s4147-018-04369-4

[48] Xing JJ, Jia YL, Peng ZR, Shi YF, He Q, Shu F, et al. Characterization of molecular identity and pathogenicity of rice blast fungus in Hunan province of China. Plant Disease. 2017;**101**:557-561

Chapter 4

A Toolbox for Managing Blast and Sheath Blight Diseases of Rice in the United States of America

Yulin Jia, Melissa H. Jia, Xueyan Wang and Haijun Zhao

Abstract

Rice blast disease caused by the fungus *Magnaporthe oryzae* and rice sheath blight disease caused by the fungus *Rhizoctonia solani* are two major hurdles for stable rice production worldwide. Presently, fungicides are still needed to manage these two devastating fungal pathogens. After two decades of research efforts, a toolbox has been assembled with the following components: (1) insight into pathogen genomic identity and pathogen avirulence (*AVR*) genes that can be used to enhance plant breeding; (2) new mapping populations and germplasm and genetic stocks that can be used as starting materials to identify effective host resistance (*R*) genes; (3) user-friendly disease evaluation methods that can be used to accelerate the identification and utilization of *R* genes; (4) validated effective *R* genes that are readily available for improving genetic resistance; (5) host genetic markers that can be used to accelerate the development of new resistant germplasms/cultivars; and (6) an improved understanding of resistance mechanisms that can facilitate the engineering of resistance in commercial varieties. Appropriate employment of these tools in breeding and crop protection will reduce production costs and create an environmentally benign, sustainable rice production system.

Keywords: resource, resistance gene, avirulence gene, interaction, innate immunity

1. Introduction

In the twentieth century, researchers around the globe focused on studying plant pathogens to develop effective pesticides and cultural practices. Since the late twentieth century, this focus has shifted to identifying resistant resources, effective resistance (*R*) genes, and deploying them in precision agricultural systems. Rice has been grown in the United States for over 300 years and is concentrated in the Southern US, including the states of Arkansas, Mississippi, Missouri, Louisiana, Texas, and California. Among them, Arkansas is located in south-central USA at ~35° N latitude, 92° W longitude and produces ~50% of the total rice production in the USA. The total annual acreage of rice in the USA is presently about 1.5 million hectares, producing about 2% of the total world rice production. Rice is being consumed domestically and/or utilized as by-products. Recently, more rice is being consumed domestically, but the majority of rice produced in the USA is exported. As a result, the USA is one of the top exporting countries in the international market. Rice production in the USA has evolved to a highly mechanized, flood intensive irrigated system with the use of airplanes, tractors, computers, lasers, fertilizers, and pesticides at its disposal. Yield per hectare is currently about 7.5 tons/hectare [1] and has been one of the top breeding priorities. Rice breeding programs in the USA are associated with private companies such as Rice Tech Inc., BASF, and major state university agriculture experiment stations, consisting of one or more rice breeders, pathologists, and other scientists. Additionally, the USDA Agriculture Research Service (ARS) has conducted research in Stuttgart, Arkansas, since 1931 [2]. Soon after the establishment of the USDA, ARS, Dale Bumpers National Rice Research Center (DB NRRC, 1998), the molecular plant pathology program has been performing translational research to tackle the major constraints of rice production.

1.1 Rice blast disease

One of the major constraints for rice production in the USA is rice blast. Blast disease of rice is caused by the filamentous fungus *Magnaporthe oryzae*

Figure 1.
Photographs showing symptoms of leaf and panicle blast and asexual spores of rice blast fungus. (A) Panicle damage caused by blast; (B) severe blast lesions on rice seedlings affecting rice seedling establishment; (C) blast lesions on a rice leaf after diseased leaf from a field was placed in a petri dish with a prewetted filter paper for 24 h; and (D) four asexual spores of the rice blast fungus [9]. The pictures were taken either with an iPhone, with a dissecting microscope, or with a Nikon eclipse microscope.

DOI: http://dx.doi.org/10.5772/intechopen.86901

(synonymous with *Pyricularia oryzae*) which belongs to the *M. grisea* species complex. The *M. grisea* species complex is known to infect a wide range of monocots causing numerous diseases. However, infection of *M. oryzae* is highly specific to its host—rice (*Oryza sativa*) [3]. The infection of a *M. oryzae* isolate to an alternative species was only demonstrated under greenhouse conditions [4]. *M. oryzae* is a polycyclic pathogen that can reproduce 3–5 generations during a single crop season depending on geographic regions [5]. *M. oryzae* can survive in debris and seeds from previous crop seasons, and the fungi carrying debris and seeds are the primary sources of inoculum for blast epidemics [6–8]. Infection of *M. oryzae* starts with asexual conidia. The conidia germinate within a few hours after attachment and penetrate the host cells. Visible symptoms on rice leaves can be seen as early as 5 days after initial contact. A single blast lesion can produce thousands of conidia within a week and these conidia can spread to another rice plant through air, dew/water, and physical contact. Each conidium is capable of causing the loss of a single rice panicle (**Figure 1**).

1.2 Sheath blight disease

The soil-borne, necrotrophic *Rhizoctonia solani* species have a wide range of host plant species. The anastomosis group AG1-IA of *R. solani* infects rice and causes sheath blight disease. *R. solani* is a monocyclic fungus. The life cycle of *R. solani* begins with mycelia growth from sclerotia soon after attachment onto rice seedlings/plants. The mycelia then move upward along the sheaths and leaves of rice plants, ultimately resulting in damages on the sheaths, leaves, and grains. The life cycle ends with the formation of overwintering structures, sclerotia on the sheaths, leaves, seeds, and in soils [10] (**Figure 2**).

Figure 2.
Photographs showing sheath blight disease on the sheaths, leaves, and grains (A and B) and young mycelia sheath blight fungus with 45 and 90° angles (C). Pictures were taken with an iPhone or with a Nikon eclipse microscope.

1.3 The epidemics, climate, and damages

In the Southern US, rice blast disease can be found annually and occasionally results in significant crop damages. However, sheath blight disease occurs more often than blast disease partially due to high-density cultivation. An extended dew period and light are known to stimulate sporulation of *M. oryzae*. Light rain is known to keep plant surfaces wet and create near 100% relative humidity, helping the attachment and penetration of the conidia of *M. oryzae*. High humid conditions also favor the growth, infection, and spread of *R. solani* to other leaves and other plants [10]. In California, there is no rain during the rice-growing season. As a result, significant yield loss due to blast has not been reported [11]. Sheath blight disease has not been reported in California either despite a phenotypically similar disease, the aggregate sheath spot of rice caused by *Rhizoctonia oryzae-sativae* [12], commonly occurring. Presently, substantial fungicides have been used to prevent crop losses of these fungal diseases in the USA.

2. Pathogen genomic identity and pathogen avirulence (*AVR*) genes

Knowledge of pathogen populations is important to identify effective *R* genes and develop long-lasting strategies to prevent crop loss due to diseases. DNA fingerprint based on MGR586, mitochondrial DNA Restriction Fragment Length Polymorphism (RFLP), mating type, vegetative compatibility, virulence, DNA sequencing, simple sequence repeat (SSR) markers, and avirulence (*AVR*) gene analyses have been interchangeably used to characterize *M. oryzae* populations [13–20]. The genetic identity of blast populations evaluated by SSR is not significantly different among rice production areas in the Southern US. The identity, however, is significantly different over the past 6 decades [19], suggesting that the environmental dynamics overtime such as weather, deployed rice varieties, and soil fertility in these years may play important roles in shaping the genetic identity of blast fungi. The pathogenicity of blast races (isolates) has been routinely evaluated with the international rice differential system since 1960s [20]. The most commonly found blast races are IB1, IB17, IB49, and IC17 while IA1, IA37, IA65, IA69, IA113, IB21, IB25, IB37, IB41, IC1, IC9, IE1k, IG1, and IH1 are the least commonly found blast races in the Southern US, whereas in California, IG1 is the only predominant blast race [19]. Similar blast races to those in the Southern US were also found in the winter nursery for the Southern US rice breeders in Puerto Rico [21].

The fungi purified from sheath blight-like diseased samples were evaluated with DNA markers, anastomosis grouping, speed of *in vitro* growth, and infection assays with detached leaf and microchamber assays [22]. All sheath blight-causing agents in 102 rice samples were determined to be *R. solani* with a diagnostic DNA marker derived from a ribosomal DNA internal transcribed spacer. Anastomosis grouping tests were conducted in cooperation with Dr. Craig Rothrock's lab (Department of Plant Pathology, University of Arkansas, Fayetteville, Arkansas, USA). A total of 13 testers, namely, (ID Al 1-4, AG-B1); (ID521, AG-9); (ID CI, AG-8); (ID1529, AG-7); (NTA3-1, AG-6); (ID ST6-1, AG-5); (ID AH-1, AG-4); (ID W14 L, AG-3); (ID RI-64, AG2-2); (ID F56 L, AG2-1); (ID M43, AG1-1C); (ID Cs-Ka, AG1-IA); and (ID SFBV-1, AG1-IB), from different hosts were used. All the 102 isolates were determined to be IG1-IA. Three groups—fast growing (such as RR0321-4, RR0319-8, RR0101-1); intermediate growing (such as RR0305, RR0316-1); and slow growing (such as RR0316-1, RR0140-1, RR0141-1)—were identified by measuring the growth of each isolate in a nutrient-supported petri dish. The speeds of growth were found to be closely correlated with the lengths of disease lesions in the detached leaves of two rice varieties, suggesting that the fast-growing isolates were more virulent than those of slow-growing isolates [22]. These characterized isolates have been used to identify genetic resistance and molecular studies ever since.

3. Mapping populations and improved rice germplasm and genetic stocks

3.1 Mapping populations

Rice germplasms with different *R* resources is a prerequisite for developing improved rice varieties with *R* genes providing overlapped resistance to various blast races (isolates). In the Southern US, tropical japonica rice varieties are mainly grown, whereas in the state of California, temperate japonica rice varieties are grown. Major resistant resources to *M. oryzae* in the Southern US are mainly from indica rice varieties such as Tetep, Te Qing, and Zhe733. Complete resistant resources to *R. solani* have not been identified; however, moderate resistance from rice germplasms such as Jasmine 85 has been identified. These resistant resources were used to develop mapping populations and adaptive germplasms through single seed descend and doubled haploid breeding strategies (**Figure 3** and **Table 1**).

3.2 Improved rice germplasms and genetic stocks

Germplasms with improved resistance to both blast and sheath blight diseases are helpful for rice breeders to develop new rice cultivars [34]. Four rice germplasms, LJRIL103 (PI 660982), LJRIL158 (PI 660983), LJRIL186 (PI 660984), and LJRIL220 (PI 660985), with resistance to both blast and sheath blight diseases were

Figure 3.
Photograph showing a view of the rice research plots of USDA ARS DBNRRC and the University of Arkansas Rice Research Center, Stuttgart, Arkansas, USA, 2016. Most rice resources and mapping populations were advanced in similar field plots. The picture was taken with a drone in 2016.

Name of genetic sources	Plant identification	Key information	Number	Year of release	Reference
C/M doubled haploid	GSOR 200001–200325	Sheath blight resistance	325	2006	[23]
Early/Katy mapping population	GSOR 100361–100600	Blast resistance	240	2007	[24]
K/Z mapping population	GSOR100001–100355	Molecular map/blast resistance	355	2007	[25]
SB5 mapping population	GSOR 101601–102,174	Blast and sheath blight resistance	574	2009	[26]
Katy//M202 backcrossing lines	GSOR 102501–102544	Blast resistance	42	2012	[27]
Weedy red rice mapping population 1	GSOR 303101–303287	Blast	187	2015	[28]
Weedy red rice mapping population 2	GSOR 303301–303536	Blast and sheath blight resistance	236	2015	[28]
USDA core collection	GSOR 310001–311795	Blast resistance	1795	2015	[29–33]

Table 1.
List of major genetic resources for blast and sheath blight resistance in the USA. Most of the rice germplasms are available at USDA-GSOR (www.ars.usda.gov/GSOR).

identified. They were identified from 800 progenies of a cross between US-adapted rice germplasm Lemont with Jasmine 85 [26]. These germplasms contain suitable agronomic traits in addition to the aromatic nature of LJRIL103, LJRIL158, and LJRIL186. Disease resistance and aromatic genes were tagged with DNA makers to ensure their incorporations.

Loss-of-function mutants can help identify the functionality of the corresponding wild-type allele [35]. For example, lesion mimic mutants (LMMs) with a phenotype resembling hypersensitive cell death without pathogen attack are useful for studying the molecular basis of plant innate immunity. A rice LMM was identified from the rice cultivar Katy after treatment with fast neutrons [36]. The severe lesion mimic phenotype of LMM1 can be induced by blast pathogens and water-related stress, respectively (M.S. Jia and Y. Jia, unpublished data). LMM1 has an enhanced resistance to both blast and sheath blight disease [36]. Genetic analysis suggests that a single recessive gene is responsible for the lesion mimic phenotype in LMM1. Further characterization of the underlying gene in LMM1 will help elucidate the mechanisms of plant innate immunity and abiotic stress responses.

The abovementioned mapping populations, characterized rice germplasms and genetic stocks, are now being used to map and clone *R* genes to both rice blast and sheath blight disease and develop DNA markers for marker-assisted breeding [37].

4. User-friendly disease evaluation methods

In the Southern US, genetic resistance to *M. oryzae* was investigated by Drs. Atkins, Johnston, and Marchetti [20, 38, 39]. Analyses of disease reactions to

Figure 4.
Photographic presentation of the massive production of the sheath blight inoculant for field evaluation. Step 1: mixing A (corn chips) and B (rye) in a 2:1 weight ratio, adding water, and autoclaving twice. Step 2: growing mycelia in petri dishes containing PDA media until the appearance of white sclerotia (C). Step 3: mixing mycelia from C with a mixture of A and B from step 1, and incubating in a sterilized plastic or metal container for 3–5 days until the appearance of white sclerotia (D and E). Step 4: air drying mycelia and sclerotia in brown bags at 24°C with a fan (F). Step 5: grinding mycelia with a grinder (G) before inoculating plants under field conditions.

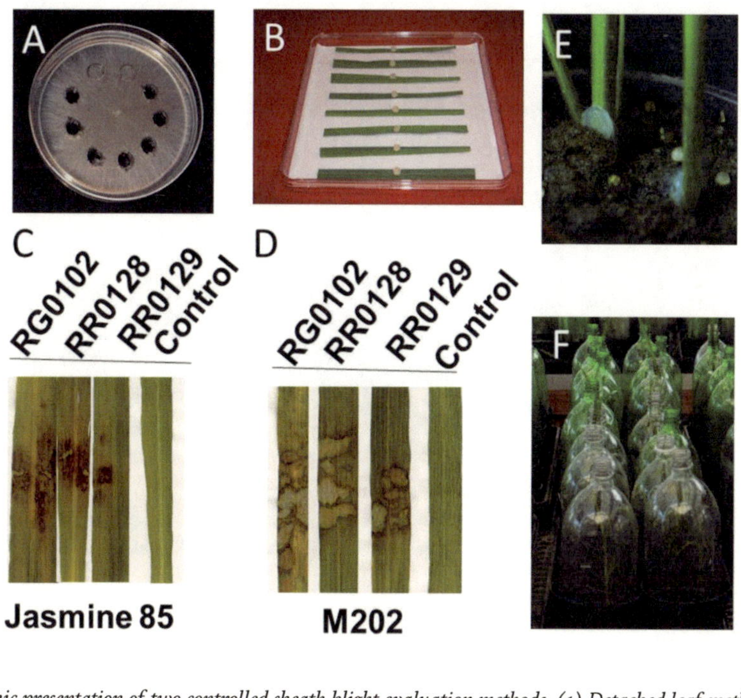

Figure 5.
Photographic presentation of two controlled sheath blight evaluation methods. (1) Detached leaf method: mycelia grown on PDA media (A), and PDA plugs removed from a were placed onto detached leaves (6–12 cm in length) (B) at 24°C for 3 days. Symptoms of detached leaves from rice varieties jasmine 85 and M202 after inoculations with three R solani *isolates versus the control PDA without pathogens (C and D). (2) Soft-drink bottle method: PDA plugs from A were placed onto the bottom of sheaths (E) and covered with 2-L soft-drink bottles (F) for 3–5 days until stable symptoms appeared. Length of lesions was measured for both methods as the severity of disease reactions.*

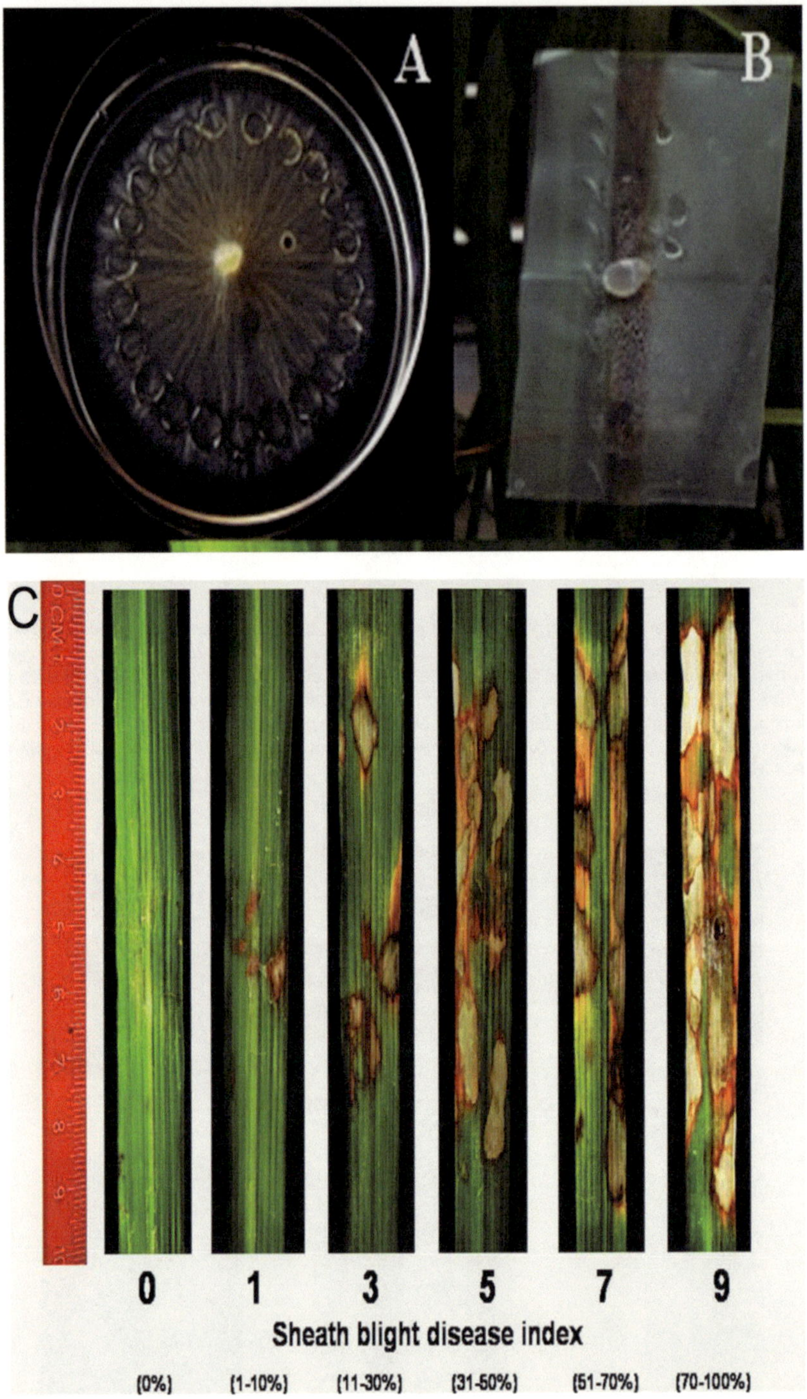

Figure 6.
Parafilm method for sheath blight disease evaluation: a PDA containing mycelia (A) in a petri dish containing mycelia after 3 days of culturing at 30°C was removed and covered with parafilm and wrapped onto the second youngest leaf for 3–5 days (B) until stable symptoms appeared. A rating scale based on visual length and area of symptoms was assigned as indicated, with 0 representing immunity and 9 representing extreme susceptibility (C).

M. oryzae have been performed under field conditions where complex biotic and abiotic factors impacting the inheritance of resistance were encountered resulting in inconsistencies of disease reactions. In 1999, Dr. Marchetti and his colleagues demonstrated that disease reactions under an upland blast nursery were reliable to identify *R* genes among breeding lines [40]. Under greenhouse conditions, the phenotypes of rice to *M. oryzae* are categorized as 0–5 where 0 represents complete immunity, 1 represents hypersensitive cell death showing tiny brown spots, 2 represents infected lesions without mycelia, and, for susceptible reactions, 3–5 exhibit different sizes of lesions with visible mycelia coincident with different levels of resistance [32]. Phenotypes evaluated under the upland rice blast nursery were verified with 200 individuals of a mapping population under greenhouse conditions at DBNRRC [41]. Since then, the greenhouse methods have been used to determine the inheritance and genetic mechanisms of blast resistance [32, 41, 42]. In 2015, several IRRI monogenic lines generously donated by IRRI were added to further identify blast *R* genes under greenhouse conditions [43].

The early evaluation of sheath blight relied on replicated field plot experiments with fungal mycelia grown in corn chips or rye (**Figure 4**).

Disease reactions were scored by visually rating the disease severity on the sheaths and leaves of whole plants. The results of the evaluations are useful for mapping *R* genes. As an alternative, greenhouse methods such as detached leaf, soft-drink bottles, and parafilm methods were developed to validate and verify the function of *R* genes (**Figures 5** and **6**). These greenhouse methods are being used routinely for initial *R* gene discovery because they use less time, labor, land, and fertilizer.

5. Effective *R* genes

5.1 Effective major *R* genes

A total of 14 known major blast *R* genes have been used in the USA since 1960s. **Table 2** lists their chromosomal locations, representing germplasms, DNA markers to monitor respective *R* genes, and the avirulent and virulent races of these selected rice germplasms (**Table 2**). Based on field observations, most blast *R* genes are dominant whereas a single haplotype of *R* gene is effective for resistance. Among them, six dominant blast *R* genes *Pia*, *Piks*, *Pi66(t)*, *Pikh*, *Pikm*, and *Pi43(t)/Pi1*, and one recessive *R* gene *pid* were on chromosome 11. Comprehensively, one was found on chromosome 2, two on chromosome 6, one on chromosome 8, one on chromosome 9, and two dominants on chromosome 12. Three of the dominant *R* genes, *Pi9*, *Pi42(t)*, and *Pi43(t)*, provide resistance to all races, while *Pita2/Ptr* is effective to all races except IE1k.

The genetic markers linked or derived from the cloned *R* genes were developed to predict resistance function and to monitor the existence of each of the *R* genes [31–33, 44–52]. Differential blast races were identified (**Table 2**) and have been used to validate their predicted resistance efficacies.

5.2 Effective minor *R* genes

Distinct phenotyping variation of rice after infection via *M. oryzae* in different rice germplasms and in the same germplasm at different growth stages under greenhouse [53] and field conditions are also referred as dilatory, partial, field, and adult resistance interchangeably [54]. A total of 11 blast *R* quantitative trait loci (QTLs) responsible for a phenotypic variation ranging from 5.17 to 26.53% were identified with different blast races under greenhouse conditions [55] (**Table 3**) and verified with different blast isolates/races [56]. Using the same method, four additional blast *R* QTLs were identified from different rice germplasms [57].

Chr.	Name of *R* gene	Selected germplasm	Marker	Name of blast races		Reference
				Avirulence	Virulence	
2	*Pi-b*	Saber, Te-Qing	RM208, Pib dom	IB1, IB45, IH1, IG1, IC17, IE1, IE1k	IB49, IB54	[31]
6	*Piz(t)*	Zenith	RM527, AP4791, AP5659-1, AP5659-5	IH1, IG1, IC17, IE1k	IA45, IB1, IB49, IB54, IB33	[32]
6	*Pi9*	IR9660-48-1-1-2 (GSOR310687)	KS6/KS28	IA45, IB1, IB49, IB54, IB45, IH1, IG1, IC17, IE1, IE1k		[33]
8	*Pi42(t)*	Zhe733	RM72	IA45, IB1, IB49, IB54, IB45, IH1, IG1, IC17, IE1, IE1k		[44]
9	*Pii*	Dawn		IH1		[39]
11	*Pia*	Bluebonnet		IB1		[39]
11	*Pikh*	Lebonnet	RM224	IB45, IB54, IH1, IG1	IB49	[45]
	Piks	M2354	E/P RM224	IB54	IA45, IB49, IB33, IB45, IH1, IG1, IC17, IE1, IE1k	[46]
11	*Pi66(t)*	DGWG		IB54	IB45, IC17, IG1, IH1	[47]
11	*Pikm*	Tsuyuake	Q/P RM224	IB45, IB54, IH1, IG1	IC17	[46]
11	*Pi43(t)/ Pi1*	Zhe733	RM1233	IA45, IB1, IB49, IB54, IB45, IH1, IG1, IC17, IE1, IE1k		[44]
11	*Pid*	Lebonnet		IB1	IA45, IB49, IB54, IB45, IH1, IG1, IC17, IE1, IE1k	[45]
12	*Pita*	Katy	YL100/YL102, YL155/YL87	IB49, IC17	IE1k	[29, 48–50]
12	*Ptr (Pita2)*	Katy	HJ16–12	IA45,IB1, IB49, IB54, IB45, IH1, IG1, IC17, IE1	IE1k	[51, 52]

Table 2.
DNA markers and resistance efficacies of deployed blast R *genes in the USA since 1960 (***Figure 7***).*

DOI: http://dx.doi.org/10.5772/intechopen.86901

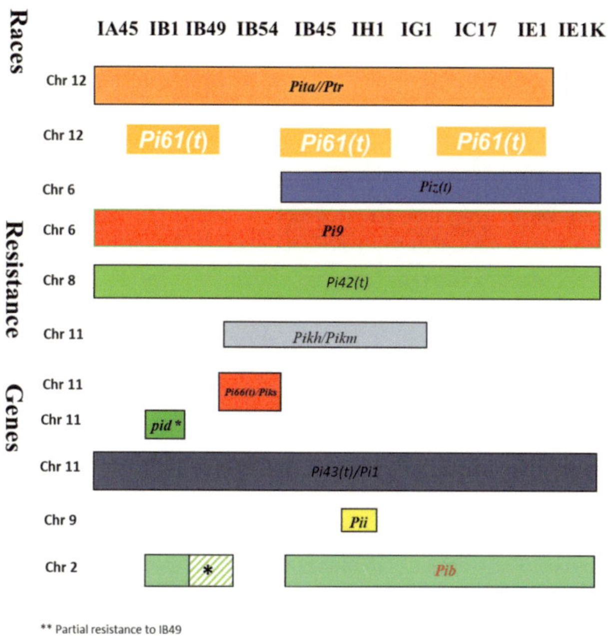

Figure 7.
Graphic presentation of resistance spectra of blast R *genes in the USA. The common races, name of* R *genes, and chromosomal locations are indicated.*

QTL	Chr.	Blast race	Marker interval	Nearest marker locus (physical location in MB)	Phenotypic variation (%)	Nearest major R genes
qBLAST 3	3	IB45	RM251–RM338	RM282 (12.4)	5.17	
qBLAST8.1	8	IB49	RM6863–RM72	RM1148 (4.0)	6.69	*Pi36*
qBLAST8.2	8	IC17	RM310–RM72	RM72 (6.8)	7.22	
qBLAST9.1	9	IB54	RM257–RM108	RM257 (17.7)	4.64	
qBLAST9.2	9	IC17	RM257–RM107	RM108 (17.9)	7.62	NBS-LRR
qBLAST9.3	9	IC17	RM107–RM245	RM215 (21.2)	4.49	
qBLAST11	11	IB45	RM206–RM224	RM224 (27.8)	26.53	*Pikm/Pik*
		IB54	RM206–RM224	RM224 (27.8)	19.6	
qBLAST12.1	12	IB1	RM6998–OSM89	OSM89 (7.9)	5.44	*Pi-ta/Ptr*
qBLAST12.2		IB49	RM247–RM277	OSM89 (7.9)	9.7	
		ID1	RM247–RM277	OSM89 (7.9)	10.18	

Chr. indicates chromosome, MB indicates megabase pair, NBS-LRR indicates the protein with nucleotide-binding sites—leucine-rich repeat domain is often encoded by the R gene.

Table 3.
List of minor resistance genes to rice blast disease with indicated nearby major R *genes and NBS-LRR proteins [55, 56].*

Thus far, major sheath blight *R* genes have not been identified. However, the major sheath blight *R* QTL *qShB9-2* responsible for 24.3–27.2% of phenotypical variation using microchamber and mist chamber assays, respectively, and other nine minor *R* QTLs to sheath blight were also identified [58, 59]. These sheath blight *R* QTLs were verified with replicated field plot experiments in multiple locations [60]. This demonstrated that there exist useful genetic factors that can be used for breeding. DNA markers linked to these *R* QTLs can not only be used to pyramid resistance into new rice varieties via marker-assisted breeding but can also be used to clone and characterize genes underlying these *R* QTLs.

6. Resistance effectiveness

M. oryzae is a hemi-biotrophic organism with an extended period of biotrophic invasion that forced the evolution of robust major blast *R* genes in host. The resistance mediated by major blast *R* genes follows the gene-for-gene model where the *R* genes in rice detect the corresponding *AVR* genes in *M. oryzae* in triggering resistance responses [61]. The existence of *AVR-Pita1* in US blast populations suggest that *AVR-Pita1* may play an important role in fitness and pathogenicity. Ironically, what is needed for pathogens to survive also makes the pathogen less virulent and fit. This never-ending booming-and-busting cycle of host-pathogen interactions presents a unique opportunity to develop durable resistance. In the Southern US, after the blast epidemics in 1980s, a blast-resistant rice variety Katy was released in 1990 [62]. Katy contains a cluster of major *R* genes at the *Pi-ta* locus from the landrace indica variety Tetep and *Piks* from tropic variety Newbonnet [41]. Further analysis of Katy revealed that there are three linked blast *R* genes, *Pi-ta* and *Pi-ta2/Ptr* genes near the centromere of rice chromosome 12. *Pi-ta* is a classical *R* gene with NBS-LRR [63] and *Ptr*, which is allelic to *Pi-ta2*, encodes a predicted protein with four armadillo repeats [52]. *Ptr* was shown to confer resistance to a wide range of blast races except for IE1k and help *Pi-ta* with unknown mechanisms [52]. To date, a handful of rice varieties with the *Pita*, *Pita2/Ptr* cluster in a linkage block including Katy, Drew, Madison, Kaybonnet, Cybonnet, Banks, Ahrent, Catahoula, and Templeton have been released in the Southern US since 1990 [64–66]. Amei and colleagues showed that the *Pi-ta* gene has been bred into cultivated species of rice for decades [67]. The counter resistance from the pathogen usually occurs after breeders release a new resistant rice variety [68]. One of the counter resistance strategies of *M. oryzae* is to alter the structural integrity and expression of the *AVR* genes. The blast races (isolates) with partial, complete deletions, point mutations altering amino acids, and transposon insertions at the *AVR-Pita1* locus have been found in commercial rice fields in the Southern US since the release of *Pi-ta* [16–18]. The resistance mediated by the *Pi-ta/Pi-ta2/Ptr* gene cluster has been stable for over two decades. Consistently, most blast populations were found to carry *AVR-Pita1* [16–18] that verified the durability of resistance mediated by *Pi-ta/Pi-ta2/Ptr*. The observed resistance durability could be due to the lack of deployment of rice cultivars with the *Pi-ta/Pi-ta2/Ptr* genes to force the loss of *AVR-Pita1*. This is consistent with the fact that limited *Pi-ta/Pi-ta2/Ptr* containing rice varieties have been grown due to moderate yield advantages compared to other rice varieties lacking the genes since their releases [69]. Alternatively, it is also fully possible that *AVR-Pita1* is important for the survival of *M. oryzae* with unknown mechanisms.

7. Summary

In the USA, any rice cultivar with one or two major blast *R* genes will continue to be effective to prevent rice blast disease. On the other hand, a combination of major

R QTLs, suitable plant architecture, and growth rate should be considered to prevent sheath blight disease. A defense gene expression and cell reaction study suggested that strong resistance responses mediated by *Pi-ta* could be initiated as early as 24 h after pathogen inoculation [70]. However, the molecular mechanisms underlying *Pi-ta* or *Ptr*-mediated disease resistance pathways [71], the interactions between major blast *R* genes and *R* QTL [74], the role of micro RNA/long noncoding RNA in rice disease resistance [72–75], and the relation of resistance versus productivity are still largely unclear [69]. Therefore, a clear understanding of the abovementioned plant innate immunity systems will be required for engineering resistance via genome editing. The lack of robust major *R* genes to *R. solani* may be due to the saprophytic nature of *R. solani* where the pathogen feed on the dead tissue of rice plants. Comparative analysis of defense genes in different hosts of *R. solani* may help identify useful *R* genes [76]. The genome of *R. solani* is mosaic and the draft sequence of *R. solani* IG1-IA genome is readily available [77]. Moving forward, the completion of whole genome sequencing will be the next urgent step to identify clues to manage *R. solani*. In brief, continued identification and characterization of *R* genes will be essential to safeguard rice crops. Ultimately, fungicides will be significantly reduced to prevent rice blast and sheath blight diseases in the future.

Acknowledgements

The authors thank Michael Lin, Tracy Bianco, Heather Box, Alan Sites, and Laduska Sells of USDA ARS DBNRRC and Mary Jia of Arkansas Schools of Math, Sciences and the Arts; Dr. Guangjie Liu of University of Arkansas Rice Research and Extension Center for excellent technical assistance; and Tyler Franzen for a photo taken by a drone. Special thanks are also given to all other scientists and supporting staff members of DB NRRC, and UA RREC for their continued support and useful discussion and interactions with the Molecular Plant Pathology program. For critical reviews we thank Drs. Trevis Huggins and Yong-Bao Pan of USDA ARS and two anonymous reviewers. USDA is an equal opportunity provider and employer.

Author details

Yulin Jia[1*], Melissa H. Jia[1], Xueyan Wang[1,2] and Haijun Zhao[1]

1 USDA ARS, Dale Bumpers National Rice Research Center, Stuttgart, Arkansas, USA

2 University of Arkansas Rice Research and Extension Center, Stuttgart, Arkansas, USA

*Address all correspondence to: yulin.jia@ars.usda.gov

References

[1] Web page Ricepedia North America. Available from: http://ricepedia.org/rice-around-the-world/north-america [Accessed: 07 December 2018]

[2] Stuttgart City Guide 2018-2019. p. 698

[3] Couch BC, Fudal I, Lebrun M-H, Tharreau D, Valent B, van Kim P, et al. Origins of host-specific populations of the blast pathogen *Magnaporthe oryzae* in crop domestication with subsequent expansion of pandemic clones on rice and weeds of rice. Genetics. 2005;**170**:613-630. DOI: 10.1534/genetics.105.041780

[4] Jia Y, Gealy D, Lin MJ, Wu L, Black H. Carolina foxtail (*Alopecurus carolinianus*): Susceptibility and suitability as an alternative host to rice blast disease (*Magnaporthe oryzae* [formerly *M. oryzae*]). Plant Disease. 2008;**92**:504-507

[5] Jia Y, Zhou E, Lee S, Bianco T. Co-evolutionary dynamics of rice blast resistance gene *Pi-ta* and *Magnaporthe oryzae* avirulence gene *AVR-Pita1*. Phytopathology. 2016;**106**:676-683

[6] Long DH, Correll JC, Lee FN, Tebeest DO. Rice blast epidmics initiated by infected rice grain on the soil surface. Plant Disease. 2001;**85**(6):612-616

[7] Ravelson H, Ramonta RI, Tharreau D, Sester M. Long-term survival of blast pathogen in infected rice residues as major source of primary inoculum in high altitude upland ecology. Plant Pathology. 2018;**67**:610-618

[8] Jia Y. A user-friendly method to isolate and single spore the fungi *Magnaporthe oryzae* and *Magnaporthe grisea* obtained from diseased field samples. Plant Health Progress. 2009;**10**:1-4. DOI: 10.1094/PHP-2009-1215-01-BR

[9] Jia Y, Gealy D. Weedy red rice has novel resistance resources to biotic stress. Crop Journal. 2018;**6**:443-450. DOI: 10.1016/j.cj.2018.07.001

[10] Uppala S, Zhou XG. Rice sheath blight. The plant health instructor. 2018. DOI: 10.1094/PHI-I-2018-0403-01. Available from: https://www.apsnet.org/edcenter/intropp/lessons/fungi/ascomycetes/Pages/RiceSheath.aspx [Accessed 14 December 2018]

[11] Greer C. What a Year for Rice Blast. Farmprogress. 2010. Available from: https://www.farmprogress.com/rice/what-year-rice-blast [Accessed 10 June 2019]

[12] Gunnell PS, Webster RK. Aggregate sheath spot of rice in California. Plant Disease. 1984;**68**:529-531

[13] Correll JC, Boza EJ, Seyran E, Cartwright RD, Jia Y, Lee FN. Examination of the rice blast pathogen population diversity in Arkansas, USA-stable or unstable? In: Wang G-L, Valent B, editors. Advances in Genetics, Genomics and Control of Rice Blast Disease. New York: Springer Sci +Businesses Media B.V.; 2009. pp. 217-228

[14] Correll JC, Harp TL, Guerber JC, Zeigler RS, Liu B, Cartwright RD, et al. Characterization of *Pyricularia grisea* in the United States using independent genetic and molecular markers. Phytopathology. 2000;**90**(12):1396-1404

[15] Li J, Li L, Jia Y, Wang Q, Fukuta Y, Li C. Characterization of field isolates of *Magnaporthe oryzae* with mating type, DNA fingerprinting, and pathogenicity assays. Plant Disease. 2016;**100**:298-303

[16] Zhou E, Jia Y, Singh P, Correll J, Lee FN. Instability of the *Magnaporthe oryzae* avirulence gene *AVR-Pita* alters

virulence. Fungal Genetics and Biology. 2007;**44**:1024-1034

[17] Dai Y, Jia Y, Correll J, Wang X, Wang Y. Diversification and evolution of the avirulence gene *AVR-Pita1* in field isolates of *Magnaporthe oryzae*. Fungal Genetics and Biology. 2010;**47**:974-980

[18] Xing J, Jia Y, Correll JC, Lee FN, Cartwright R, Cao M, et al. Analysis of genetic and molecular identity among field isolates of the rice blast fungus with an international differential system, rep-PCR, and DNA sequencing. Plant Disease. 2013;**97**:491-495

[19] Wang X, Jia Y, Wamishe Y, Jia MH, Valent B. Dynamic changes in the rice blast population in the USA over six decades. Molecular Plant-Microbe Interactions. 2017;**30**:803-812. DOI: 10.1094/MPMI-04-17-0101-R

[20] Atkins JG, Robert AL, Adair CR, Goto K, Kozaka T, Yanagida R, et al. An international set of rice varieties for differntiaing races of *Piricularia oryzae*. Phytopathology. 1967;**57**:297-301

[21] Jia Y, McClung AM. First report of multiple races of the rice blast fungus *Magnaporthe oryzae* in Puerto Rico. Plant Disease. 2016;**10**(6):1242. DOI: 10.1094/PDIS-12-15-1391-PDN

[22] Wamishe Y, Jia Y, Singh P, Cartwright RD. Identification of field isolates of *Rhizoctonia solani* to detect quantitative resistance in rice under greenhouse conditions. Frontiers of Agriculture in China. 2007;**1**:361-367

[23] Chu QR, Linscombe SD, Rush MC, Groth DE, Oard J, Sha X, et al. Registration of a C/M doubled haploid mapping population of rice. Crop Science. 2006;**46**:1416

[24] Jia Y, Moldenhauer KAK. Development of monogenic and digenic rice lines for blast resistance genes *Pi-ta, Pi-kh/Pi-ks*. Journal of Plant Registrations. 2010;**4**:163-166

[25] Liu G, Bernhardt J, Jia MH, Wamishe Y, Jia Y. Molecular characterization of the recombinant inbred line population derived from a japonica-indica rice cross. Euphytica. 2008;**159**:73-82

[26] Jia Y, Liu G, Jia MH, McClung AM. Registration of a rice gene mapping population of Lemont × jasmine 85 recombinant inbred lines. Journal of Plant Registrations. 2015;**9**:128-132

[27] Jia Y, Berger G, McClung AM. Registration of 42 blast resistant medium grain rice genetic stocks with suitable agronomic, yield, milling yield, and grain characteristics. Journal of Plant Registrations. 2016;**10**:316-324

[28] Liu Y, Qi X, Gealy DR, Olsen KM, Caicedo AL, Jia Y. QTLs analysis for resistance to blast disease in US weedy rice. Molecular Plant-Microbe Interactions. 2015;**28**:834-844. DOI: 10.1094/MPMI-12-14-0386-R

[29] Wang X, Fjellstrom R, Jia Y, Yan WG, Jia MH, Scheffler BE, et al. Characterization of *Pi-ta* blast resistance gene in an international rice core collection. Plant Breeding. 2010;**129**:491-501

[30] Liu Y, Jia Y, Gealy DR, Goad DM, Caicedo AL, Olsen KM. Marker development for rice blast resistance gene *Pi66(t)* and application in USDA rice mini-core collection. Crop Science. 2016;**56**:1-8. DOI: 10.2135/cropsci2015.07.0422

[31] Roychowdhury M, Jia Y, Jia MH, Fjellstrom B, Cartwright R. Identification of the rice blast resistance gene *Pi-b* in the national small grains collection. Phytopathology. 2012;**102**:700-706

[32] RoyChowdhury M, Jia Y, Jackson A, Jia MH, Fjellstrom R, Cartwright R. Analysis of rice blast resistance gene *Pi-z* using pathogenicity assays and DNA markers. Euphytica. 2012;**184**:35-47

[33] Scheuermann K, Jia Y. Identification of a *Pi9*-containing rice germplasm with a newly developed robust marker. Phytopathology. 2016;**100**(2):298-303

[34] Jia Y, Liu G, Correa-Victoria FJ, McClung AM, Oard JH, Bryant RJ, et al. Registration of four rice germplasm lines with improved resistance to sheath blight and blast diseases. Journal of Plant Registrations. 2012;**6**:95-100

[35] Jia Y. Understanding the molecular mechanisms of rice blast resistance using rice Mmutants. In: Shu QY, editor. Induced Plant Mutations in the Genomics Era. Rome: Food and Agriculture Organization of the United Nations; 2009. pp. 375-378

[36] Jia Y. Registration of lesion mimic mutant of Katy rice. Crop Science. 2005;**45**:1675

[37] Jia Y. Marker assisted selection for the control of rice blast disease. Pesticide Outlook. 2003;**14**:150-152

[38] Atkins JG, Johnston TH. Inheritance in rice of reaction to races 1 and 6 of *Piricularia oryzae*. Phytopathology. 1965;**55**:993-995

[39] Marchetti MA. Race-specific and rate-reducing resistance to rice blast in US rice cultivars. In: Zeigler RS, Leong SA, Teng PS, editors. Rice Blast Disease. Wallingford, UK: CAB International; 1994. pp. 231-244

[40] Lai XH, Marchetti MA, Petersen HD. Comparative slow-blasting in rice grown under upland and flooded blast nursery culture. Plant Disease. 1999;**83**:681-684

[41] Jia Y, Lee F, McClung A. Determination of resistance spectra to US races of *Magnaporthe oryzae* causing blast in a recombinant inbred line population. Plant Disease. 2009;**93**:639-644

[42] Jia Y, Valent B, Lee FN. Determination of host responses to *Magnaporthe grisea* on detached rice leaves using a spot inoculation method. Plant Disease. 2003;**87**:129-133

[43] Wang J, Correll JC, Jia Y. Characterization of rice blast resistance genes in rice germplasm with monogenic lines and pathogenicity assays. Crop Protection. 2015;**72**:132-138

[44] Lee S, Wamishe Y, Jia Y, Liu G. Identification of two major resistance genes against race IE-1k of *Magnaporthe oryzae* in the indica rice cultivar Zhe733. Molecular Breeding. 2009;**24**:127-134

[45] Marchetti MA, Lai X, Bollich CN. Inheritance of resistance to *Pyriculaira oryzae* rice cultivars grown in the United States. Phytopathology. 1987;**77**:799-804

[46] Costanzo S, Jia Y. Sequence variation at the rice blast resistance gene *Pi-km* locus: Implications for the development of allele specific markers. Plant Science. 2010;**178**:523-530

[47] Liu Y, Qi X, Young ND, Olsen KM, Caicedo AL, Jia Y. Characterization of resistance genes to rice blast fungus *Magnaporthe oryzae* in a "green revolution" rice variety. Molecular Breeding. 2015;**35**:52. DOI: 10.1007/s11032-015-0256-y

[48] Jia Y, Wang Z, Fjellstrom RG, Moldenhauer KAK, Azam MA, Correll J, et al. Rice *Pi-ta* gene confers resistance to the major pathotypes of the rice blast fungus in the US. Phytopathology. 2004;**94**:296-301

[49] Jia Y, Wang Z, Singh P. Development of dominant rice blast *Pi-ta* resistance

gene markers. Crop Science. 2002;**42**:2145-2149

[50] Jia Y, Redus M, Wang Z, Rutger JN. Development of a SNLP marker from the *Pi-ta* blast resistance gene by tri-primer PCR. Euphytica. 2004;**138**:97-105

[51] Jia Y, Martin R. Identification of a new locus, *Ptr(t)* required for rice blast resistance gene *Pi-ta*-mediated resistance. Molecular Plant-Microbe Interactions. 2008;**21**:396-403

[52] Zhao H, Wang X, Jia Y, Minkenberg B, Wheatley M, Fan J, et al. The rice blast resistance gene *Ptr* encodes an atypical protein and confers broad spectrum disease resistance. Nature Communications. 2018;**9**:2039. DOI: 10.1038/s4147-018-04369-4

[53] Chen X, Jia Y, Wu B. Evaluation of rice responses to blast fungus *Magnaporthe oryzae* at different growth stages. Plant Disease. 2018;**103**:132-136. DOI: 10.1094/PDIS-12-17-1873-+-RE

[54] Marchetti MA. Dilatory blast resistance in rice lines exotic to the southern United States. Plant Disease. 1983;**67**:1362-1364

[55] Jia Y, Liu G. Mapping quantitative trait loci for resistance to rice blast. Phytopathology. 2011;**101**:176-187

[56] Xing J, Jia MH, Correll JC, Yuan L, Deng H, Jia Y. Confirming and identifying new loci for resistance to rice blast disease using field isolates of *Magnaporthe oryzae* in the U.S. Crop Science. 2015;**55**:1-8

[57] Yang H, Jia MH, Jia Y, Xing J, Venu R, Bellizzi M, et al. Molecular mapping of four blast resistance genes using recombinant inbred lines of 93-11 and Nipponbare. Journal of Plant Biology. 20131;**56**:91-97

[58] Liu G, Jia Y, Correa-Victoria F, Prado GA, Yeater KM, McClung A, et al. Mapping quantitative trait loci responsible for resistance to sheath blight in rice. Phytopathology. 2009;**99**:1078-1084

[59] Nelson J, Oard JH, Groth D, Utomo H, Jia Y, Liu G, et al. Sheath-blight resistance QTLs and in japonica rice germplasm. Euphytica. 2011;**184**:23-34

[60] Liu G, Jia Y, McClung A, Oard JH, Lee FN, Correll JC. Confirming QTLs and finding additional loci responsible for resistance to rice sheath blight disease. Plant Disease. 2013;**97**:113-117

[61] Jia Y, McAdams SA, Bryan GT, Hershey H, Valent B. Direct interaction of resistance gene and avirulence gene products confers rice blast resistance. European Molecular Biology Organization Journal. 2000;**19**:4004-4014

[62] Moldenhauer KAK, Lee FN, Norman RJ, Helms RS, Well RH, Dilday RH, et al. Registration of 'Katy' rice. Crop Science. 1990;**30**:747-748

[63] Bryan GT, Wu K-S, Farrall L, Jia Y, Hershey HP, McAdams S, et al. A single amino acid difference distinguishes resistant and susceptible alleles of the rice blast resistance gene Pi-ta. The Plant Cell. 2000;**12**:2033-2045

[64] Wang X, Lee S, Wang J, Ma J, Bianco TA, Jia Y. In: Rice WY, Bao J, editors. Current Advances on Genetic Resistance to Rice Blast Disease. London, UK: IntechOpen Limited; 2014. pp. 195-217. DOI: 10.5772/56824. Available from: http://www.intechopen.com

[65] Jia Y. Artificial introgression of a large chromosome fragment around the rice blast resistance gene *Pi-ta* in backcross progeny and several elite rice cultivars. Heredity. 2009;**103**:333-339

[66] Jia Y, Jia MH, Wang X, Liu G. Indica and japonica crosses resulting in linkage block and recombination suppression

on rice chromosome 12. PLoS ONE. 2012;**7**(10):e43066

[67] Amei A, Lee S, Mysore KS, Jia Y. Statistical inference of selection and divergence of rice blast resistance gene Pi-ta. Genes, Genomes, Genetics. 2014;**4**:2425-2432

[68] Dean R, Van Kan JAL, Pretorius ZA, Hammond-Kosack KE, Di Pietro A, Spanu PD, et al. The top 10 fungal pathogens in molecular plant pathology. Molecular Plant Pathology. 2012;**13**:414-430. DOI: 10.1111/j.1364-3703.2011.00783.x

[69] Wang X, Jia MH, Ghai P, Lee FN, Jia Y. Genome-wide association of rice blast resistance and yield related components of rice. Molecular Plant-Microbe Interactions. 2015;**28**:1383-1392

[70] Wang Z, Lin H, Valent B, Rutger JN, Jia Y. Cytological and molecular analyses of disease resistance to the rice blast fungus. Chinese Journal of Rice Science. 2007;**21**:335-340

[71] Costanzo S, Jia Y. Alternatively spliced transcripts of *Pi-ta* blast resistance gene in *Oryza sativa*. Plant Science. 2009;**177**:468-478

[72] Chen X, Jia Y, Jia MH, Pinson S, Wang X, Wu B-W. Functional interactions between major rice blast resistance genes, *Pi-ta* and *Pi-b*, and minor blast resistance QTL. Phytopathology. 2018;**108**:1095-1103. DOI: 10.1094/PHYTO-02-18-0032-R

[73] Venu RC, Jia Y, Gowda M, Jia MH, Jantasuriyarat C, Stahlberg E, et al. RL-SAGE and microarray analysis of the rice transcriptome after *Rhizoctonia solani* infection. Molecular Genetics and Genomics. 2007;**278**:421-431

[74] Li W, Jia Y, Liu F, Wang F, Fan F, Wang J, et al. Genome-wide identification and characterization of long non-coding RNAs responsible to *Dickeya zeae* in rice. RSC Advances. 2018;**8**:34408-34417

[75] Li W, Jia Y, Liu F, Wang F, Fan F, Wang J, et al. Integration analysis of small RNA and degradome sequencing reveals microRNAs responsive to *Dickeya zeae* in resistant rice. Molecular Science. 2019;**20**:222. DOI: 10.3390/ijms20010222

[76] Rioux R, Manmathan H, Singh P, Reyes B, Jia Y, Tavantzis S. Comparative analysis of putative pathogenesis-related gene expression in two *Rhizoctonia solani* pathosystems. Current Genetics. 2011;**57**:391-408

[77] Nadarajah K, Razali NM, Cheah BH, Sahruna NS, Ismail S, Tathode M, et al. Draft genome sequence of *Rhizoctonia solani* anastomosis group 1 subgroup 1A strain 1802/KB isolated from rice. Genome Announcements. 2017;**5**(43):e01188-e01117. DOI: 10.1128/genomeA.01188-17

Chapter 5

Evaluation of Resistance of US Rice Breeding Lines to the Rice Blast Pathogen

Chunda Feng and James C. Correll

Abstract

Rice blast, caused by the fungus *Magnaporthe oryzae* (anamorph: *Pyricularia oryzae*), is a ubiquitous disease that threatens rice production in the USA and worldwide. Growing resistant cultivars is the most economical and effective way to manage this disease. Multiple races exist in the *M. oryzae* population in the USA. It is necessary to know the resistance spectrum of rice cultivars to the prevalent rice blast races in the areas where they are grown. Twelve isolates of *M. oryzae* collected from the southern US rice-growing region were used in this study. The genetic diversity of these isolates was evaluated with genetic and molecular methods, and the pathogenicity to different rice blast resistance genes was determined by the disease reaction of two sets of near-isogenic lines containing one blast *R* gene per line. From 2005 to 2016, about 200 Uniform Regional Rice Nursery (URRN) breeding lines have been tested with 9–12 reference isolates annually, and a total of 2377 breeding lines have been tested. The varieties with good resistance to rice blast disease have been identified. The results could be useful for the management of rice blast disease in the southern US rice production area.

Keywords: rice, blast disease, avirulence gene, resistance gene, breeding lines

1. Introduction

Rice is one of the most important staple food crops worldwide, feeding over half of the world's population [1]. The demand for rice continues to increase with the increase in the global population. The USA grows approximately 1.5 million hectares of rice annually and produces about 8–11 million metric tons of rice valued at 3.6 billion dollars (**Figure 1**) [2] . Although the USA is a relatively small rice producer accounting for less than 2% of the total rice production worldwide, it is a major rice exporter that occupies 6–13%, with an average of 10%, of the world rice export market (**Figure 1**), making the USA one of the top rice exporters in the world [3].

Rice blast disease, caused by the fungus *Magnaporthe oryzae* (anamorph: *Pyricularia oryzae*), is one of the most important diseases on rice worldwide and is responsible for approximately 30% of rice production losses globally [1, 4]. A wide range of management practices have been used to reduce losses from rice blast. For example, cultural practices such as crop rotation, controlling the timing and amount of nitrogen applied, and managing the flood depth in the field may reduce the impact of blast [5]. A number of fungicides also are effective in managing rice

IntechOpen

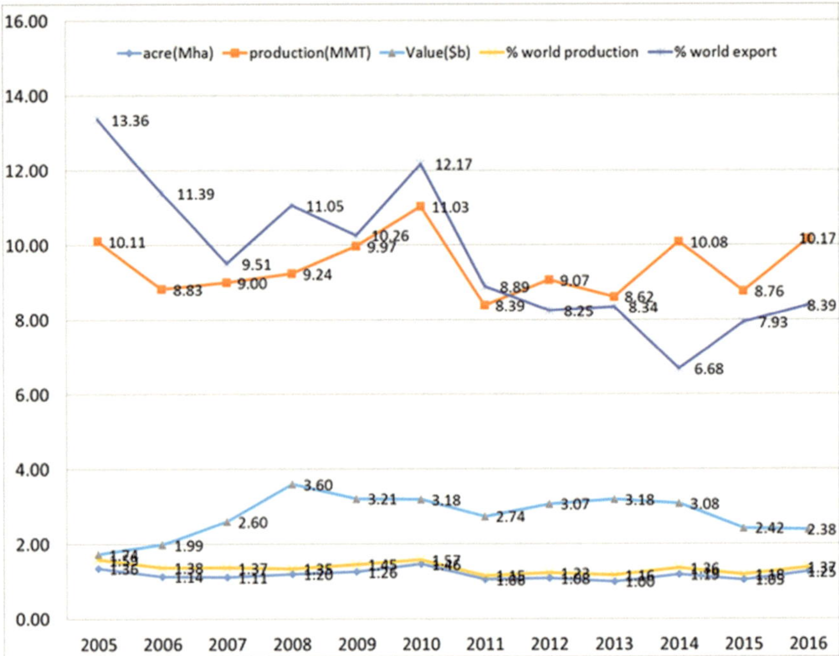

Figure 1.
Rice production in the USA and its percentages of world total rice production and export. Mha, million hectare; MMT, million metric tons.

blast disease [4]; however, it is not a preferred management option due to environmental concerns and cost. Growing resistant cultivars is the most economical and effective way to manage this disease [4, 6]. Many rice blast *R* genes have been characterized, some of which have been widely used in rice breeding programs worldwide [6–8]. The *R* genes recognize the corresponding specific avirulence genes from the pathogen and initiate defense mechanism [9]. For example, the *R* gene *Pita* can interact with the counterpart *AVR-Pita* from the pathogen and confer resistance [10]. However, the changes in avirulence genes can result in the loss of function of the corresponding *R* genes. For example, the *R* gene *Pita* has deployed in rice cultivars in the southern USA and provided durable resistance for a long period of time [11], but the resistance of the *Pita* gene was overcome by race IE1k in 2004 [12].

The population of *M. oryzae* in the southern USA has been intensively studied [13–18]. Multiple races exist in the *M. oryzae* population in the USA. For example, race IB49 and IC17 were the most prevalent races in Arkansas [13–15], with occasional epidemics due to race IE1k or "race K" type isolates [12]. Near-isogenic lines, each containing a targeted blast resistant gene, in either a Japonica-type variety Lijiangxingtuanheigu (LTH) background [19] or Indica-type CO39 background [20], have been used for race identification in Asia [21]. In the USA, the *M. oryzae* population has been intensively studied [13–18, 22], but the relationship between races to individual rice blast *R* genes in the USA is largely unknown [22]. In addition, it is necessary to evaluate the resistance spectrum of newly developed rice breeding lines to the prevalent rice blast races in the southern US rice-growing region before they are released.

The objective of this study was to summarize the disease reactions of a wide range of rice germplasm from the Uniform Regional Rice Nursery (URRN) lines to 12 reference isolates of the rice blast pathogen from 2005 to 2016.

2. Diversities of the 12 US reference isolates of *M. oryzae*

2.1 Twelve US reference isolates of *M. oryzae*

Twelve isolates of *M. oryzae*, collected from the southern USA, were used as reference isolates to test the URRN lines during 2005–2016 (**Table 1**). Among them, six isolates (49D, #24, A119, A264, A598, IB33) were collected from AR; four isolates (TM2, ID13, ZN7, and ZN15) were collected from TX; one isolate, IB54, from LA; and one isolate, ZN46, from FL (**Table 1**). These isolates represented 10 races, including IB49 (49D, A119, and A598), IB33, IB54, IC1 (ZN46), IC17 (A264), ID13, IE1 (ZN7), IE1k (TM2), and IG1 (#24). Most isolates were used each year on a different set of URRN lines. Isolate IB33 has been tested in 11 years but not in 2007. Isolate IB54 has not been tested until 2009. Isolate ID13 has been tested in 7 years, but not in 2005, 2007, 2009, 2013, and 2014.

2.2 Genetic diversity of the 12 reference isolates

The genetic diversity of the 12 reference isolates was evaluated by vegetative compatibility analysis [13] and molecular methods. Vegetative compatibility analysis indicated that three isolates A598, ZN15, and ZN46 belonged to vegetative compatibility group (VCG) US-01; isolates TM2, #24, and A264 belonged to VCG US-02; two isolates 49D and A119 belonged to VCG US-03; and other two isolates IB33 and IB54 belonged to VCG US-04 (**Table 1**). The VCG of isolate ID13 was not determined.

Using Pot 2 primers [23], the repetitive element-based polymerase chain reaction (Rep-PCR) was used to DNA fingerprint the 12 reference isolates. The amplicon patterns of 49D, IB33, and IB54 based on Pot 2 primers were identical; TM2 and ZN7 were identical to each other; isolate 24 and A264 were identical to each other, but they had one extra band compared to that of TM2 and ZN7; ZN15 and ZN46 had similar patterns (**Figure 2**). The mating types of these isolates were determined by using mating-type-specific primers [24]. The results suggested that six isolates, 49D,

Isolate	Vegetative compatibility group (VCG)	Mating type	RACE	Year	Origin
49D	US-03	I	IB49	1985	AR
TM2	US-02	II	IE1K	2004	TX
#24	US-02	II	IG1	1992	AR
A119	US-03	II	IB49	1992	AR
A264	US-02	II	IC17	1993	AR
A598	US-01	I	IB49	1992	AR
IB33	US-04	I	IB33		AR
IB54	US-04	I	IB54	1959	LA
ID13		II	ID13	1982	TX
ZN7	US-02	II	IE1	1995	TX
ZN15	US-01	I	IB1	1996	TX
ZN46	US-01	I	IC1	1996	FL

Table 1.
Background information on the 12 US reference isolates of M. oryzae used in this study.

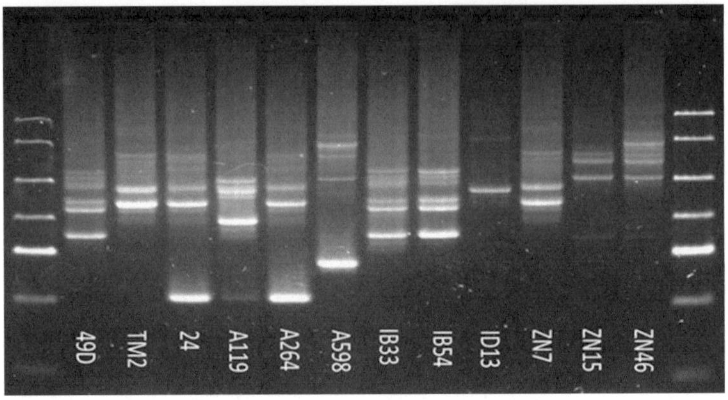

Figure 2.
Rep-PCR band patterns of 12 reference isolates amplified with Pot 2 primers.

A598, IB33, IB54, ZN15, and ZN46, belonged to mating type I, while other six isolates TM2, #24, A119, A264, ID13, and ZN7 belonged to mating type II (**Figure 3**).

Seven avirulence genes were assessed using specific primers to each gene (**Table 2**) [25–28]. The entire *AVR-Pita* fragment could be amplified from nine isolates with primers YL149/YL169, but not from isolates TM2, IB33, and ID1. The coding regions of the avirulence gene *AVR-Pib* was found in all 12 reference isolates (amplified with the *AVR-Pib* F3/R3 primers); however, the promoter region of the *AVR-Pib* gene (amplified with the *AVR-Pib* F2/R2 primers) was not found in isolates 49D, IB33, and IB54. The avirulence gene *AVR-Pikm* was only found in four isolates, 49D, IB33, IB54, and ID13. The other four avirulence genes, *AVR-CO39*, AVR-Pi9, *AVR-Pikz*, and *AVR-Piz-t,* were present in all 12 reference isolates (**Figure 4**).

2.3 Testing the US reference isolates on IRRI near-isogenic rice lines

2.3.1 IRRI rice blast near-isogenic lines

The 12 US reference isolates were tested on 31 LTH NILs (containing 24 blast *R* genes) and 20 CO39 NILs (containing 14 *R* genes) in three independent tests, with two replications in each test. Two cultivars, M204 and Francis, were included as susceptible controls.

2.3.2 Inoculation of blast pathogen and disease screening

Rice seed was planted in plastic trays filled with river sand mixed with potting soil in the greenhouse at the University of Arkansas, Fayetteville, AR, USA. Iron

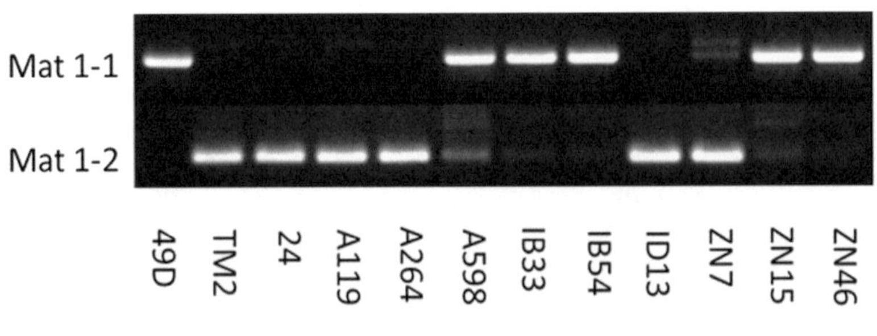

Figure 3.
Mating type analysis of the 12 reference blast isolates.

Target gene	Primer name	Sequences
AVR-Pita	YL149	TGACCGCGATTCCCTCCATT
	YL169	CGACCGTTTCCGCC
AVR1-CO39	AVR1-CO39F1	GATCTGTAAATTACATA
	AVR1-CO39R1	GGATCCGCCGTCGCCTCC
AVR-Pi9	AVRPi9F	CTG CTC CAT CTT GTT TGG CC
	AVRPi9R	CAC TAG TAC AAG CAC TAA CC
AVR-Pib	AVR-PibF2	TGGAGAAGACTTTGATGC
	AVR-PibR2	TAGTTGCCATTATGCGTTC
	AVR-PibF3	ATGCGTTCCTCAACCACTTT
	AVR-PibR3	TTATTCCACGGTATATTTGCTGCC
AVR-Pikz	AVR-PikzF	TGACGCAGCTTGAGTTGT
	AVR-PikzR	TCCGAGCAATCAACTCTG
AVR-Pikm	AVR-PikmF	TTATCGCCCCTATATTGC
	AVR-PikmR	TTATCGCCCCAACACGGA
AVR-Piz-t	AVR-Piz-tF	ATGCAGTTCTCAACCATC
	AVR-Piz-tR	CTATTGGCGCTGAGCCT

Table 2.
Primers used to amplify seven avirulence genes from the 12 US reference isolates of M. oryzae.

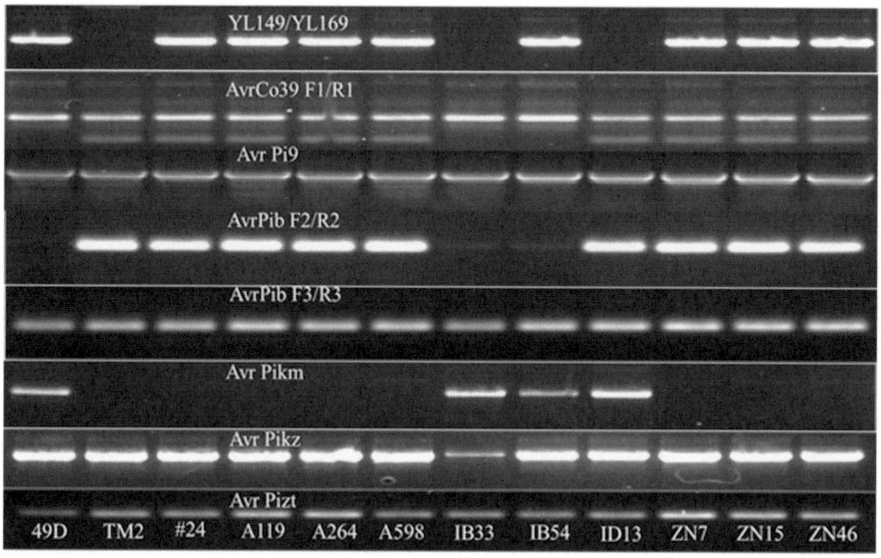

Figure 4.
Detection of seven avirulence genes in the 12 US reference isolates of M. oryzae.

sulfate was applied to the newly emerged seedlings. The plants were fertilized with Miracle-Gro All-Purpose Plant Food 20-20-20 once a week during each test. Plants were inoculated approximately 14–20 days after planting. Each isolate was grown on rice bran agar (RBA) [13] for approximately 7–10 days and then reinoculated on new RBA plates for 7–10 days. Spores were collected in cold water and adjusted to a concentration of 200,000 spores/ml per isolate. Each tray was inoculated with 50 ml of inoculum mixed with 0.02% Tween 20 with an air compressor sprayer. After inoculation, the plants were incubated at 100% relative humidity in a mist chamber at approximately 22°C for 24 h, allowed to dry for 2–3 h before being moved to the greenhouse. The inoculated plants were incubated in the greenhouse for 6 days. On the 7th day after inoculation, 15–20 plants of each line were scored according to a standard 0–9 disease rating scale developed by IRRI [22]. Lines rated 0 to 3 were considered resistant, whereas those rated 4–9 were considered susceptible.

2.3.3 Virulence of US reference isolates of M. oryzae on IRRI rice NILs

Among the blast reference isolates, IB33, 49D, and TM2 were the most virulent isolates, whereby only 8–13 (15.7–21.6%) of the NILs were resistant. Isolates IB54, ID13, and #24 were the least virulent isolates with 33–38 (64.7–74.5%) of the NILs being resistant. Other isolates were intermediate in virulence (**Table 3**).

NILs containing gene *Pi3(t)* were susceptible to all reference isolates tested. NILs containing *R* genes *Pia*, *Pi19(t)*, *Pii*, *Piks*, and *Pizt*, were only resistant to one isolate. Those lines containing *Pikh*, *Pikp*, and *Pita* were resistant to two isolates. NILs containing the *R* genes, *Pi1*, *Pi7(t)*, *Pik*, *Pikp*, *Pikm*, *Pit*, and *Piz*, were resistant to three isolates. NILs containing *Pi9(t)* or *Pi12(t)* were resistant to all isolates (**Table 4**). Lines containing genes *Pib*, *Pi11(t)*, and *Pita-2/Ptr* [29] were resistant to 9 or 11 isolates.

Line	A598	ZN15	ZN46	24	A264	ZN7	TM2	49D	A119	IB33	IB54	ID13
R	16	21	17	33	31	15	13	11	27	8	38	34
%	31.4	41.2	33.3	64.7	60.8	29.4	25.5	21.6	52.9	15.7	74.5	66.7
S	35	30	34	18	20	36	38	40	24	43	13	17
%	68.6	58.8	66.7	35.3	39.2	70.6	74.5	78.4	47.1	84.3	25.5	33.3

Table 3.
Number and percentage of rice NILs resistant or susceptible to blast isolates.

ID	Gene	49D	TM2	24	A119	A264	A598	IB33	IB54	ID13	ZN7	ZN15	ZN46
		IB49	IE1K	IG1	IB49	IC17	IB49	IB33	IB54	ID13	IE1	IB1	IC1
IRBLA-A	*Pia*	S	S	S	S	S	S	S	R	S	S	S	S
IRBLA-C	*Pia*	S	S	S	S	S	S	S	R	S	S	S	S
IRBLI-F5	*Pii*	S	S	S	S	S	S	R	R	S	S	S	S
IRBLKS-F5	*Piks*	S	S	S	S	S	S	S	R	S	S	S	S
IRBLKS-S	*Piks*	S	S	S	S	S	S	R	R	S	S	S	S
IRBLK KA	*Pik*	S	S	R	S	S	S	S	R	R	S	S	S
IRBLKP-K60	*Pikp*	S	S	R	S	S	S	S	S	R	S	S	S
IRBLKH-K3	*Pikh*	S	S	R	S	S	S	S	S	R	S	S	S
IRBLZ FU	*Piz*	S	R	S	S	R	S	S	S	R	S	S	S
IRBLZ 5-CA	*Piz-5*	R	S	S	R	R	R	R	R	R	R	R	S
IRBLZT-T	*Piz-t*	S	S	S	S	S	S	S	R	S	S	S	S
IRBLTA CT 2	*Pita*	S	S	R	S	R	S	S	S	S	S	S	S
IRBLB-B	*Pib*	S	R	R	R	R	R	S	S	R	R	R	R
IRBLT-K59	*Pit*	S	S	S	R	R	S	S	R	S	S	S	S
IRBLSH-S	*Pish*	R	R	R	R	R	R	R	R	R	R	R	R
IRBLSH-B	*Pish*	S	S	R	R	R	R	S	R	S	S	R	S

ID	Gene	49D	TM2	24	A119	A264	A598	IB33	IB54	ID13	ZN7	ZN15	ZN46
		IB49	IE1K	IG1	IB49	IC17	IB49	IB33	IB54	ID13	IE1	IB1	IC1
IRBL 1-CL	*Pi1*	S	S	R	S	S	S	S	R	R	S	S	S
IRBL 3-CP 4	*Pi3*	S	S	S	S	S	S	S	S	S	S	S	S
IRBL 5-M	*Pi5(t)*	S	S	R	R	R	S	S	R	R	S	R	S
IRBL 7-M	*Pi7(t)*	S	S	R	R	R	S	S	S	R	S	S	S
IRBL 9-W	*Pi9*	R	R	R	R	R	R	R	R	R	R	R	R
IRBL 12-M	*Pi12(t)*	R	R	R	R	R	R	R	R	R	R	R	R
IRBL 19-A	*Pi19(t)*	S	S	S	S	R	S	S	S	S	S	S	S
IRBLKM TS	*Pikm*	S	S	R	S	S	S	S	R	R	S	S	S
IRBL 20-IR 24	*Pi20*	S	S	S	R	R	R	S	R	R	S	R	R
IRBLTA 2-PI	*Pita-2*	R	S	R	R	R	R	S	R	R	R	R	R
IRBLTA 2-RE	*Pita-2*	R	S	R	R	R	R	S	R	R	R	R	R
IRBLTA CP 1	*Pita*	S	S	R	S	S	S	S	R	S	S	S	S
IRBL 11-ZH	*Pi11(t)*	S	R	R	R	R	R	S	R	R	R	R	R
IRBLZ 5-CA (R)	*Piz-5*	R	R	S	R	R	S	R	R	R	S	R	S
LJXTHG	*LTH*	S	S	S	R	R	S	S	R	S	S	S	S
IR 85430	*Pish*	S	R	R	R	R	S	S	R	R	S	R	R
IR 85424	*Pish*	S	R	R	R	R	S	S	R	R	S	R	R
IR 93322	*Pish*	S	R	R	R	R	S	S	R	R	S	R	R
IR 85417	*Pib*	S	R	R	R	R	R	S	R	R	R	R	R
IR 85427	*Piz-5*	R	S	S	R	R	R	R	R	R	S	R	R
IR 85429	*Piz-t*	S	R	R	R	R	R	S	R	R	R	R	R
IR 85413	*Pi5(t)*	S	R	S	R	R	S	S	R	R	S	S	S
IR 85423	*Piks*	S	S	S	S	S	S	S	R	S	S	S	S
IR 85420	*Pik*	S	S	R	S	S	S	S	R	R	S	S	S
IR 85419	*Pik*	S	S	R	S	S	S	S	R	R	S	S	S
IR 85421	*Pikm*	S	S	R	S	S	S	S	R	R	S	S	S
IR 85422	*Pikp*	S	S	R	S	S	S	S	R	R	S	S	S
IR 85411	*Pi1*	S	S	R	S	S	S	S	R	R	S	S	S
IR 85414	*Pi7(t)*	S	S	R	S	R	S	S	R	R	S	S	S
IR 85426	*Pita*	S	S	R	R	R	S	S	R	S	R	S	S
IR 93324	*Pita*	S	S	R	R	R	S	S	R	S	R	S	S
IR 93323	*Pita-2*	R	S	R	R	R	R	R	R	R	R	R	R
IR 85425	*Pita-2*	R	S	R	R	R	R	S	R	R	R	R	R
IR 93325	*Pita-2*	R	S	R	R	R	R	S	R	R	R	R	R
CO39	*Pia*	S	S	S	S	S	S	S	S	S	S	S	S

Table 4.
Disease responses of rice NILs to US Magnaporthe oryzae reference isolates.

Four loci provided resistance to reference isolate 49D (race IB49) or IB33 (race IB33), 7 loci were resistant to isolate TM2 (race k), and 14, 16, and 17 loci were resistant to isolate IB54, isolate #24 (race IG1), and isolate ID13, respectively.

Discrepancy in disease reactions was observed among lines putatively containing the same target *R* genes. One NIL (IRBLSH-S), containing *Pish*, was resistant to all blast reference isolates, while four *Pish* containing lines were only resistant to six to eight isolates. Both NILs IRBLZT-T (in LTH background) and IR 85429 (in CO39 background) contain *R* gene *Pizt*. NIL IRBLZT-T was resistant to one isolate, while IR 85429 was resistant to 10 isolates. These discrepancies may have resulted due to a number of reasons including linkage drag from different donor parents. The *R* genes in the NILs also need to be confirmed with specific molecular markers.

Thus, NILs containing *Pia*, *Pi3*, *Pi19(t)*, and *Pi12(t)* were not useful for differentiating races of the US reference isolates tested. Resistance loci *Pi9(t)*, *Pi12(t)*, *Pib*, *Pi11(t)*, and *Pita-2* were the most effective *R* genes to the panel of US reference isolates evaluated and could be exploited to improve resistance to rice blast disease in the USA.

3. Evaluation of resistance of the US rice breeding lines to reference isolates of *M. oryzae*

3.1 Uniform Regional Rice Nursery (URRN) breeding lines

About 200 rice breeding lines, developed by the rice breeders from Arkansas, Louisiana, Mississippi, and Texas, were subjected to annual disease evaluations to the reference blast isolates at the University of Arkansas, Fayetteville, AR, in addition to the evaluation of yield and agronomic traits at various locations. A total of over 2000 breeding lines were tested during 2005–2016. The rice cultivars M204 and Francis were included in each test as the susceptible controls. The inoculation and disease scoring procedures were as described previously.

3.2 Pathogenicity of the reference isolates on the URRN lines

The susceptible control Francis was susceptible to all 12 isolates, while M204 was susceptible to 11 isolates but resistant to isolate IB54. Each year, in each test, the two susceptible controls consistently showed susceptible disease reactions with the disease rating scores ranging from 4 to 9, respectively.

The percentage of breeding lines resistant to each isolate in each year was quantified (**Figure 5**). The isolate IB33 was the most virulent isolate out of the 12 reference isolates tested, with 71.2–98% of the lines evaluated as susceptible for the 11 years tested. Overall 1963 out of 2177 lines tested (90.2%) were susceptible to IB33. Isolate 49D (race IB49) was highly virulent, with 70–90% of the lines tested susceptible for 10 of the 12 years examined. In 2007, however, 21.5% in 2007 and 48.7% in 2016 were evaluated as susceptible. Out of the 2377 lines tested in 12 years, 1673 lines (70.4%) were susceptible to 49D. Isolate TM2 (IE1k) was also considered highly virulent. In 2014 and 2015, about 75% of the lines were susceptible to TM2. In other years, over than 50% of the lines were susceptible. The lines tested in 2007 had the lowest percentage (33.5%) of susceptibility to this isolate. Overall, 1361 out of 2377 breeding lines (57.3%) were susceptible to TM2. Three isolates ID13, IB54, and #24 (IG1) were the least virulent; the percentages of susceptible breeding lines ranged between 7.5–26.7, 13.5–27.5, and 10.5–28.3%, respectively. Overall, the percentages of susceptible breeding lines to these three isolates were 18.7, 21.0, and

19.2%. The other six isolates were intermediately virulence on the lines tested with 40 to 50% of breeding lines were susceptible. In 2006, over 80% of the breeding lines were susceptible to isolate A119 (race IB49), but in the following years, only 25 to 50% of lines were susceptible to this isolate. In 12 years, 970 out of 2377 breeding lines (40.8%) were susceptible to A119.

3.3 Disease reaction of US rice breeding lines to the 12 reference isolates

All 12 reference isolates have been tested in 2010–2013 and 2016. In these 5 years, there were 45 lines that were rated as completely resistant to all isolates, and 101 lines only susceptible to one isolate. In 2010, there were 10 lines resistant to all isolates, 11 lines only susceptible to IB33, and 1 each only susceptible to TM2 and IB54. A total of 20 lines had no resistance to the 12 isolates. There were 14 lines from the 2011 set of germplasm that were resistant to all 12 isolates, 11 lines only susceptible to IB33, 1 only susceptible to TM2, and 2 only susceptible to 49D, and 8 lines susceptible to all 12 isolates. Five lines tested in 2012 were resistant to all 12 isolates, 12 lines were only infected by IB33, 1 and 3 lines were only susceptible to ID13 or 49D, respectively, while 7 lines were susceptible to all 12 lines. In 2013, 14 lines were evaluated as resistant to all isolates, 13 lines were only susceptible to IB33, 1 each only susceptible to ID13 or ZN7, 4 each only susceptible to 49D or TM2, while 5 lines were susceptible to all 12 isolates. Out of the 200 URRN lines tested in 2016, only 2 lines were resistant to all 12 isolates, 1 line only susceptible to TM2, 2 lines only susceptible to 49D, 32 lines were only susceptible to IB33, and 4 lines were susceptible to all 12 lines. In 2006 and 2008, 11 isolates were tested, but not IB54, 2 and 6 lines were resistant, and 28 and 13 lines were susceptible to all 11 isolates, respectively. In 2009, 2014, and 2015, isolate ID13 was not tested, but other 11 isolates were. There were 3, 1, and 4 lines resistant to and 11, 14 and 9 lines susceptible to all 11 isolates. In 2005, both IB54 and ID13 were not tested. No variety was found to be resistant to all 10 isolates tested. There were nine lines only susceptible to one isolate, six of them were susceptible to isolate IB33, and one each susceptible to 49D, TM2, and A598. There were 19 lines susceptible to all 10 isolates. Nine isolates were tested in 2007, but not IB33, IB54, and ID13. There were 60 lines resistant and 4 lines susceptible to all 9 isolates tested in 2007.

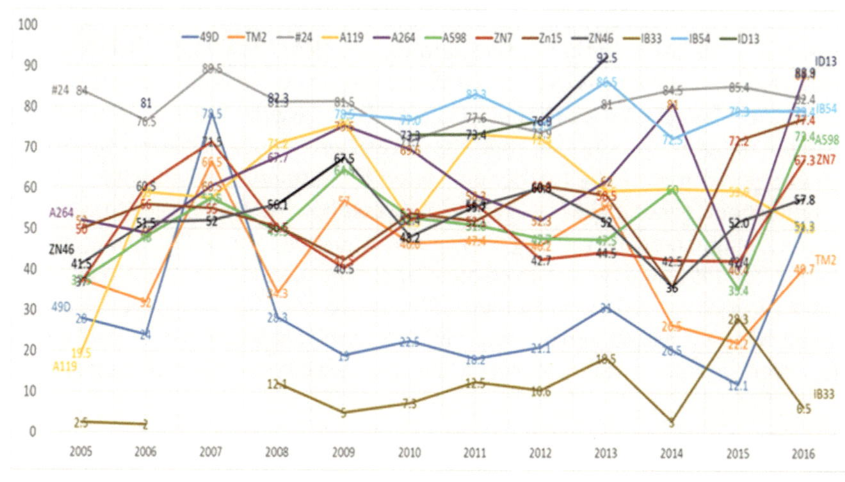

Figure 5.
Percentage of breeding lines resistant to each of the 12 reference US isolates of M. oryzae in each year from 2005 to 2016.

4. Discussion

Growing resistant cultivars has been demonstrated to be the most economical and effective way to manage rice blast disease. During 2005 to 2016, 2377 breeding lines were evaluated for disease resistance to the 12 reference isolates. Breeding lines resistant to all isolates have been found in each year of the period except 2005. Some lines were only susceptible to the most virulent isolate IB33. The use of the lines that have the broadest level of resistance to the spectrum of reference isolates would reduce the loss due to rice blast disease.

Based on the international differential cultivars and nomenclature, isolates A119, A598, and 49D are classified as race IB49 [18]. The disease reactions of many NILs tested to these three isolates were identical. However, these three isolates can be differentiated by some NILs (*R* genes), as *Pib*, *Pi11(t)*, and *Pi20* containing lines were resistant to A119 and A598 but susceptible to 49D; *Pi5(t)* and *Pit* containing lines were resistant to A119 but susceptible to 49D and A598. These results indicated that a set of differential cultivars should be chosen to more clearly demarcate races within the US rice blast pathogen population.

Any mutation, insertion, or deletion of the avirulence genes in the pathogen could cause the changes in its pathogenicity, thus resulting in the loss of function of the corresponding *R* gene and disease development. The coding region of *AVR-Pib* was found in all 12 reference isolates, but the promoter region was not amplified from isolates 49D, IB33, and IB54, and this may explain why the *Pib* gene containing line IRBLB-B cannot provide resistance to these three isolates. Some of the avirulence genes in the US population of *M. oryzae* have been studied [17, 18, 25]. However, the variation of other avirulence genes in the US population of *M. oryzae* needs to be evaluated.

Specific primers were used to detect the presence/absence of seven avirulence genes. Three avirulence genes, *AVR1-CO39*, *AVR-Pi9*, *AVR-Pikz*, *AVR-Pizt*, were present in all 12 reference isolates. According to the gene-for-gene concept [9], the corresponding *R* genes *Pi-CO39* line, *Pi9*, *Pikz*, and *Piz-t* would interact with these avirulence genes and initiate the defense response. It is unknown how many avirulence gene/*R* gene pairs could be involved in the resistance recognition process. When *AVR-Pita*1 was introduced into strains that were virulent on *Pita* containing cultivars, those transformed strains lost their pathogenicity on *Pita* containing cultivars [30], suggesting that one *R* gene recognized one corresponding avirulence gene to initiate the resistance response. If this is the case, then cultivar CO39 and lines carrying *Pi9*, *Pikz*, and *Piz-t*, would have broad-spectrum resistance to the US isolates. It has been shown that Pi9 containing line IRBL 9-W had resistance to all 12 reference isolates, but this is in contrast to the results of the NILs carrying *Pi-CO39*, *Pikz*, and *Piz-t* based on the reference isolates. If these avirulence genes in the reference isolates are functional, then the critical avirulence gene or combination of avirulence genes needs to be further evaluated for managing the disease.

A number of *R* genes to the blast pathogen disease have been identified from rice [4, 6–8]. Although more than 20 *R* genes were incorporated into the NILs, only *Pi9*, *Pi11(t)*, *Pi12(t)*, *Pib*, and *Pita-2* showed broad spectrum of resistance to the reference isolates of *M. oryzae* found in the southern USA. The *R* gene *Pita-2* has been widely used in US rice breeding programs, and has been effective, but incorporation of other *R* genes to develop more durable resistant cultivars will help to reduce the impact of rice blast disease.

5. Conclusions

The population of *M. oryzae* in the southern USA is very diverse. Breeding lines with broad spectrum of resistance to the reference isolates have been developed, and incorporation of other *R* genes to develop more durable resistant cultivars will help to reduce the impact of rice blast disease.

Acknowledgements

This research has financial support through the rice checkoff funds administered by the Arkansas Rice Research and Promotion Board.

Conflict of interest

No conflict of interest.

Author details

Chunda Feng and James C. Correll
University of Arkansas, Fayetteville, AR, USA

*Address all correspondence to: cfeng@uark.edu

References

[1] Nalley L, Tsiboe F, Durand-Morat A, Shew A, Thoma G. Economic and environmental impact of rice blast pathogen (*Magnaporthe oryzae*) alleviation in the United States. PLoS One. 2016;**11**(12):e0167295

[2] https://www.nass.usda.gov/Statistics_by_Subject/result.php?F3B539C7-F1DA-34A7-981A-053B4000EA6C&sector=CROPS&group=FIELD%20CROPS&comm=RICE

[3] https://www.statista.com/statistics/255947/top-rice-exporting-countries-worldwide-2011/

[4] Skamnioti P, Gurr SJ. Against the grain: Safeguarding rice from rice blast disease. Trends in Biotechnology. 2009;**27**(3):141-150

[5] TeBeest DO, Guerber C, Ditmore M. Rice blast. Plant Health Instructor. 2007. Available from: http://www.apsnet.org/edcenter/intropp/lessons/fungi/ascomycetes/Pages/RiceBlast.aspx

[6] Khush GS, Jena KK. Current status and future prospects for research on blast resistance in rice (*Oryza sativa L.*). In: Wang G-L, Valent B, editors. Advances in Genetics, Genomics and Control of Rice Blast Disease. Berlin, Germany: Springer Science+Business Media B.V.; 2009. pp. 1-10

[7] Srivastava D, Shamim M, Kumar M, Mishra A, Pandey P, Kumar D, et al. Current status of conventional and molecular interventions for blast resistance in rice. Rice Science. 2017;**24**(6):299-321

[8] Koide Y, Kobayashi N, Xu DH, Fukuta Y. Resistance genes and selection DNA markers for blast disease in rice (*Oryza sativa L.*). Jarq-Japan Agricultural Research Quarterly. 2009;**43**(4):255-280

[9] Flor HH. Current status of the gene-for-gene concept. Annual Review of Phytopathology. 1971;**9**:275-296

[10] Jia Y, McAdams SA, Bryan GT, Hershey HP, Valent B. Direct interaction of resistance gene and avirulence gene products confers rice blast resistance. EMBO Journal. 2000;**19**(15):4004-4014

[11] Jia Y, Wang X, Costanzo S, Lee S. Understanding the Co-evolution of the rice blast resistance gene PI-TA and Magnaporthe oryzae avirulence gene AVR-PITA. In: Wang G-L, Valent B, editors. Advances in Genetics, Genomics and Control of Rice Blast Disease. Berlin, Germany: Springer Science+Business Media B.V.; 2009. pp. 137-147

[12] Lee FN, Cartwright RD, Jia Y, Correll JC. Field resistance expressed when the *Pi-ta* gene is compromised by *Magnaporthe oryzae*. In: Wang G-L, Valent B, editors. Advances in Genetics, Genomics and Control of Rice Blast Disease. Berlin, Germany: Springer Science+Business Media B.V.; 2009. pp. 281-290

[13] Correll JC, Harp TL, Guerber JC, Zeigler RS, Liu B, Cartwright RD, et al. Characterization of *Pyricularia grisea* in the United States using independent genetic and molecular markers. Phytopathology. 2000;**90**(12):1396-1404

[14] Xia JQ, Correll JC, Lee FN, Ross WJ, Rhoads DD. Regional population diversity of *Pyricularia grisea* in Arkansas and the influence of host selection. Plant Disease. 2000;**84**(8):877-884

[15] Wang XY, Jia YL, Wamishe Y, Jia MH, Valenta B. Dynamic changes in the rice blast population in the United States over six decades. Molecular Plant-Microbe Interactions. 2017;**30**(10):803-812

[16] Zhai L. Genotypic and Phenotypic Diversity of *Pyricularia oryzae* in the Contemporary Rice Blast Pathogen Population in Arkansas. Fayetteville, AR, USA: University of Arkansas; 2009

[17] Rotich F. Rice Blast Disease in the U.S. and Africa: Determination of Pathogen Diversity and the Identification of Resistance Genes for Disease Management. Fayetteville, AR, USA: University of Arkansas; 2015

[18] Boza E. Rice Blast Disease: Pathogen Diversity, Breeding for Resistance and Variation in an Avirulence Gene (*AVR-Pita*). Fayetteville, AR, USA: University of Arkansas; 2005

[19] Tsunematsu H, Yanoria MJT, Ebron LA, Hayashi N, Ando I, Kato H, et al. Development of monogenic lines of rice for blast resistance. Breeding Science. 2000;**50**(3):229-234

[20] Telebanco-Yanoria MJ, Koide Y, Fukuta Y, Imbe T, Tsunematsu H, Kato H, et al. A set of near-isogenic lines of Indica-type rice variety CO 39 as differential varieties for blast resistance. Molecular Breeding. 2011;**27**(3):357-373

[21] Kawasaki-Tanaka A, Hayashi N, Yanagihara S, Fukuta Y. Diversity and distribution of rice blast (*Pyricularia oryzae Cavara*) races in Japan. Plant Disease. 2016;**100**(4):816-823

[22] Correll JC, Boza EJ, Seyran E, Cartwright RD, Jia Y, Lee FN. Examination of the rice blast pathogen population diversity in Arkansas, USA-Stable or unstable? In: Wang G-L, Valent B, editors. Advances in Genetics, Genomics and Control of Rice Blast Disease. Berlin, Germany: Springer Science + Business Media B.V.; 2009. pp. 217-228

[23] George MLC, Nelson RJ, Zeigler RS, Leung H. Rapid population analysis of *Magnaporthe grisea* by using rep-PCR and endogenous repetitive DNA sequences. Phytopathology. 1998;**88**(3):223-229

[24] Xu J-R, Hamer JE. Assessment of *Magnaporthe grisea* mating type by spore PCR. Fungal Genetics Newsletter. 1995;**42**:80

[25] Zhou EX, Jia YL, Singh P, Correll JC, Lee FN. Instability of the *Magnaporthe oryzae* avirulence gene *AVR-Pita* alters virulence. Fungal Genetics and Biology. 2007;**44**(10):1024-1034

[26] Zhang SL, Wang L, Wu WH, He LY, Yang XF, Pan QH. Function and evolution of *Magnaporthe oryzae* avirulence gene *AvrPib* responding to the rice blast resistance gene *Pib*. Scientific Reports. 2015;**5**:11642

[27] Farman ML, Leong SA. Chromosome walking to the *AVR1-CO39* avirulence gene of *Magnaporthe grisea*: Discrepancy between the physical and genetic maps. Genetics. 1998;**150**(3):1049-1058

[28] Jia Y, Wamishe YA, Zhou B. An expedited method for isolation of DNA for PCR from *Magnaporthe oryzae* stored on filter paper. The Crop Journal. 2014;**2**:267-271

[29] Zhao HJ, Wang XY, Jia YL, Minkenberg B, Wheatley M, Fan JB, et al. The rice blast resistance gene *Ptr* encodes an atypical protein required for broad-spectrum disease resistance. Nature Communications. 2018;**9**:2039

[30] Dai Y, Winston E, Correll JC, Jia Y. Induction of avirulence by *AVR-Pita1* in virulent U.S. field isolates of *Magnaporthe oryzae*. The Crop Journal. 2014;**2**:1-9

Chapter 6

Sustainable Strategies for Managing Bacterial Panicle Blight in Rice

Xin-Gen Zhou

Abstract

Bacterial panicle blight (BPB) is present in more than 18 countries and has become a global disease in rice. BPB is highly destructive and can cause significant losses of up to 75% in yield and milling quality. BPB is caused by *Burkholderia glumae* or *B. gladioli*, with the former being the primary cause of the disease. Outbreaks of BPB are triggered by conditions of high temperatures in combination with high relative humidity at heading. The disease cycle starts with primary infections from infected seed, soil, and irrigation water, and subsequent secondary infections result from rain splash and panicle contact. Limited management options are available for control of BPB. There are only several cultivars including hybrids with partial resistance available currently. Twelve quantitative trait loci (QTLs) associated with the partial resistance have been identified. Oxolinic acid is an effective antibacterial compound for control of BPB in Japan, but it is not labeled for use on rice in the USA and many other countries. Sustainable control of BPB relies on integrated use of available management strategies of exclusion, genetic resistance, chemical control, biocontrol, and cultural practice. Developing and use of resistant cultivars is the best strategy to minimize the damage caused by BPB and maximize rice production in the long term.

Keywords: rice, bacterial panicle blight, seedling rot, grain rot, QTLs, *Pseudomonas glumae*, *Burkholderia glumae*, *Burkholderia gladioli*, disease control, IPM, epidemiology, genetic resistance, chemical control, biocontrol, cultural practice

1. Introduction

Bacterial panicle blight (BPB), caused primarily by *Burkholderia glumae*, has become a threat to rice production globally. BPB has the potential to cause significant losses in grain yield and milling quality in epidemic years. The disease causes several types of damage, including seedling blight, sheath rot, floret sterility, grains not filling or aborted, and milling quality reduction, resulting in a reduction of yield by up to 75% [1–4]. In Japan, BPB has become one of the major rice diseases. Severe outbreaks of this disease occurred on more than 69,000 ha in 2013 and 30,000 ha in 2015 [5, 6]. In the USA, BPB has recently become as one of the most important diseases in rice in terms of economic importance. A survey found that the disease was present in approximately 60% of Louisiana rice fields [7]. In the Southern USA, significant yield losses from BPB were reported in 1995, 1996, 1998, 2000, 2010, and 2011 [1, 8–11]. In Louisiana, yield losses for severely infected fields reached 40% in

1995 and 1998 [1, 8]. In Arkansas, BPB was so severe in 2010 that yield losses were estimated at 50% in susceptible cultivars [9]. In Texas, the outbreaks of BPB resulted in an estimate of 10–20% yield loss in the Texas Rice Belt in 2010 [10, 11]. Outbreaks of this disease also occurred in rice under organic production systems in 2010 in Texas [11]. In the disease-yield loss field study, we found BPB was highly destructive and could cause yield losses ranging from 1 to 59% (83–4883 kg/ha), with yield loss increasing approximately 5% (455 kg/ha) for every unit increase in BPB severity on the rating scale of 0–9 [12]. Based on annual rice production in the Mid-South USA in 2003–2013, it is estimated that BPB caused $61 million USD of damage that would feed 1.1 million people annually (Aaron Shew, personal communication).

Effective management of BPB is critical to minimizing the damage caused by the disease and maximizing production returns. However, limited options for management of the disease are available currently. No single genes or quantitative trait loci (QTLs) for complete resistance to BPB have been found so far [13, 14]. Only a few rice cultivars with partial resistance are available for commercial use. No chemical control options are available in the USA although oxolinic acid has been used as a major control measure for BPB in Japan for more than two decades [15]. Resistant populations of *B. glumae* to oxolinic acid have been found [16–19], which limits increasing use of this antibiotic compound for management of BPB. Oxolinic acid is not labeled for use on rice in the USA and many other countries. Compared to extensive research and significant advances made on management of sheath blight caused by *Rhizoctonia solani* and rice blast caused by *Magnaporthe oryzae*, very limited research has been conducted on the development of effective and sustainable management options for control of BPB.

In this article, we focus on the review of recent advances on the development of management strategies for BPB, including exclusion, genetic resistance, chemical control, biological control, and cultural practice. In addition, world distribution of the pathogen, characteristic symptoms of BPB, and current understanding of epidemics of BPB are also included. Two review articles covering the pathogenesis of *B. glumae* and the detection of BPB have been published previously [20, 21]. The terms “BPB” and “grain rot” have been used interchangeably in the literature. However, BPB has been commonly used in the USA and Latin America, while grain rot in Japan and other countries [20]. The term BPB is used in this review article.

2. Pathogens

Since the first description of *Burkholderia glumae* (formerly *Pseudomonas glumae* Kurita and Tabei) as the bacterial pathogen causing rice seedling rot and grain rot in Japan in 1955 [22], BPB has been reported in more than 18 countries distributed in Africa, Asia, Latin America, and North America (**Table 1**). The total rice production from these countries accounted for more than 65% of total world rice production in 2018 [23]. BPB has become an increasingly important global disease in rice. In addition to *B. glumae*, *B. gladioli* has also been identified as another bacterial pathogen causing the BPB disease. Infection with *B. gladioli* produces the same symptoms as infection with *B. glumae*. The disease caused by *B. gladioli* has been reported in Arkansas (USA), China Japan, Louisiana (USA), Panama, and the Philippines, where *B. glumae* is also co-present (**Table 1**). In the USA, the cause of the BPB was not known at the time when epidemics of BPB occurred in 1995. In 1996–1997, however, when evaluating bacterial isolates from rice tissue for their ability to control the rice sheath blight fungus *R. solani*, investigators in Louisiana accidentally found that some of the *B. glumae* isolates caused panicle blighting symptoms when greenhouse grown rice plants were spay inoculated [44]. This led to the discovery of *B. glumae* as the causal agent of the BPB disease.

Country	Year	BPB pathogen	Reference
Japan	1955	*B. glumae*	[22, 24]
Taiwan (China)	1983	*B. glume*	[25]
Columbia	1989	*B. glumae*	[26]
Latin America	1989	B. glumae	[26]
Vietnam	1993	*B. glumae*	[27]
Japan	1996	*B. gladioli*	[28, 29]
The Philippines	1996	*B. glume* and *B. gladioli*	[30–32]
Louisiana (USA)	2001	*B. glume* and B. gladioli	[1, 33]
Korea	2003	*B. glumae*	[34]
China	2007	*B. glumae*	[35]
Panama	2007	*B. glume* and *B. gladioli*	[36]
Nicaragua	2008	*B. glumae*	[37]
Arkansas (USA)	2009	*B. glume* and *B. gladioli*	[1, 9]
Mississippi (USA)	2009	*B. glumae*	[1]
Texas (USA)	2009	B. glumae	[1, 10]
Honduras	2011	*B. glumae*	Lex Ceamer, personal communication
Mississippi (USA)	2012	*B. gladioli*	[38]
Costa Rica	2014	*B. glumae*	[39]
Ecuador	2014	*B. glumae*	[40]
South Africa	2014	*B. glumae*	[41]
India	2015	*B. glumae*	[42]
China	2018	*B. gladioli*	[43]

Table 1.
Countries reported with the presence of bacterial panicle blight (BPB) caused by Burkholderia glumae and B. gladioli in rice as of January 2019.

BPB of rice can be caused by either *B. glumae* or *B. gladioli*. However, the former is the primary cause of the disease. The study of Nandakumar et al. [1] found that 76 and 5% of the bacterial strains collected were *B. glumae* and *B. gladioli*, respectively. In a field survey conducted in Mississippi using PCR analysis, it was found that 84% of rice panicle samples collected were positive for *B. glumae* and 12% of the samples positive for *B. gladioli* [38]. In a recent survey conducted in nine rice-producing counties of Arkansas, all 45 virulent bacterial isolates studied were *B. glumae*, and no *B. gladioli* isolates were identified [9]. In addition, the *B. glumae* pathogen tends to be more virulent and causes more damage to rice plants when compared to the *B. gladioli* pathogen [20, 33].

3. Symptoms

The symptoms of BPB include seedling blight, sheath rot, and panicle blighting [1–4]. These symptoms can be induced by either *B. glume* or *B. gladioli*. Virulent bacterial strains produce the yellow-pigmented toxin toxoflavin on King's B agar medium (**Figure 1**), while avirulent strains do not produce this toxin [1]. Production of toxoflavin is an essential factor to induce the development of the symptoms on rice seedlings and grains [34, 45, 46].

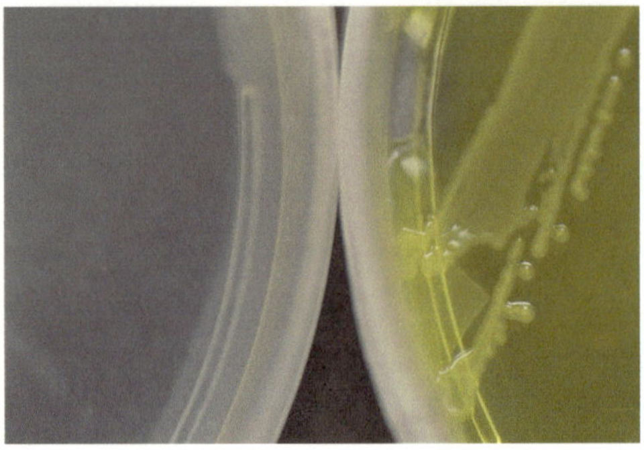

Figure 1.
Colonies of Burkholderia glumae and production of yellow pigment (toxoflavin) by B. glumae on King's B agar plate (right) vs. no pigment production control plate (left). Photo was taken at 3 days after inoculation at 30°C.

Figure 2.
A focal pattern of bacterial panicle blight (BPB) on the Presidio (cv) rice panicles (center) in a research plot inoculated with Burkholderia glumae at Beaumont, Texas.

Unlike rice sheath blight and blast, BPB is difficult to be diagnosed based on the symptoms on panicles. Similar symptoms on panicles can be caused by many abiotic and biotic factors including heat, insect damage, and secondary microorganisms [3, 4, 47]. However, BPB has the symptoms that can be distinguished from other causes. BPB occurs sporadically on individual plants or in circular or oval patterns in the field (**Figures 2** and **3**). In contrast, common panicle blanking, caused by abiotic stress such as from excessive heat, develops in the field more uniformly and does not form apparent foci. There are three important characteristics of BPB that separate it from other panicle disorders: (1) BPB often does not appear to prevent successful pollination although it can affect individual glumes or whole panicles (**Figure 4**). Thus, seed may be present on the panicle unlike panicle sterility that is caused by heat stress. (2) Infected florets initially have discoloration ranging from light green to light brown on the basal portion of the glumes with a reddish-brown margin separating this area from the rest that becomes straw-colored later (**Figures 4** and 5). (3) The rachis or branches of the panicle remain green for a while at the base of each floret, even after the glumes desiccate and turn tan (**Figures 4** and 5). Florets at the latest stages of infection usually appear to be gray or black due to the abundant growth of saprophytic fungi on the surface (**Figure 5**). The disease can cause linear lesions on sheaths with a distinct reddish-brown border and a gray and necrotic center, resulting in sheath rot (**Figure 6A**) and stem rot (**Figure 6B**). On the leaves, lesions are circular to oval with a smooth reddish-brown border and a gray or straw-colored center (**Figure 6C**). If the infected plants are young, this disease can cause seedling blighting (**Figure 6D**) or seeding rot. The symptoms of seedling rot were

Figure 3.
Symptoms of bacterial panicle blight (BPB) on a Presidio (cv) rice panicle head (arrow) in the field inoculated with Burkholderia glumae at the flowering stage at Beaumont, Texas.

Figure 4.
A close look at the symptoms of bacterial panicle blight (BPB) on Presidio (cv) rice panicles. Photo was taken approximately 2 weeks after inoculation with Burkholderia glumae at the flowering stage at Beaumont, Texas.

first reported in Japan [22] and frequently occur in young rice plants. However, these symptoms on leaves, sheaths, stems, and seedlings are rarely observed under the field conditions in the Southern USA [4]. This is one of the reasons why no scouting methods have been developed to detect and predict the development of BPB based on the symptoms on leaves and sheaths at the early crop growth stages.

Figure 5.
Comparison of the developmental symptoms of bacterial panicle blight (BPB) on infected kernels of rice (lower row) and healthy kernels (upper row). Photo was taken for rice kernels collected from different Presidio (cv) rice plants inoculated with Burkholderia glumae at the flowering stage in the field. Note the occurrence of secondary fungal infection on the discolored kernel at the late BPB development stage (lower right end).

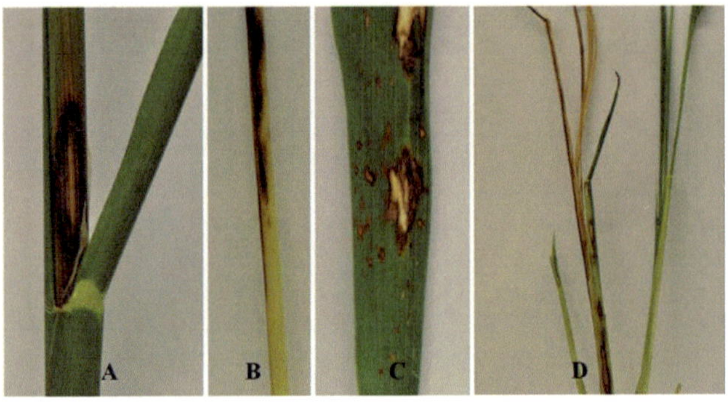

Figure 6.
Symptoms of sheath rot (A), stem rot (B), leaf lesions (C), and seedling blighting (D) caused by Burkholderia glumae in Presidio (cv) rice. Rice seedlings were inoculated with B. glumae and maintained in the greenhouse.

4. Epidemiology

The disease cycle and epidemiology of BPB of rice are not completely understood. Both *B. glumae* and *B. gladioli* species have been identified as the cause of the BPB disease. However, the former has much wider distribution in the world as shown in **Table 1**. The bacteria of both species were also found to be widely present in rice seed lots in the studies conducted in China, Japan, the Philippines, and the USA [21, 32, 48]. Therefore, infected seeds serve as the primary source of inoculum [1]. In addition, Jeong et al. [34] reported that *B. glumae* could also infect other plant species, including tomato, sesame, perilla (an herb), eggplant, and hot pepper. The bacteria are capable of inhabiting surface plants and soils under a wide range of environments [49, 50]. In a field survey conducted in Mississippi using PCR analysis, it was found that 83% of soil samples were positive for *B. glumae* and 2% of the soil samples positive for *B. gladioli* [38]. This survey also found that 85% of field irrigation water samples collected were positive for *B. glumae* and 2% of the water samples positive for *B. gladioli*. Therefore, soil and irrigation water can also serve as the sources of inoculum for the spread and development of BPB.

The bacterial pathogen invades germinated seeds, inhabits the roots and lower sheaths, and moves up the growing plant as an epiphyte (an organism growing on a plant surface, but not as a parasite) [2, 51, 52]. A recent study, using real-time fluorescence quantitative PCR to monitor the infection process of *B. glumae*, finds that the bacterium also can directly infect the rice plant by colonizing the vascular bundle of lateral roots and then spreading to upper tissues such as leaf sheaths and leaf blades through vascular system [53]. Infection by the bacterium occurs at flowering by invading rice spikelets through stomata or wound in the epidermis of glumes. The bacterium colonizes and multiplies in spikelets quickly after invasion by utilizing intermediate sugars in developing grains [51, 52]. The bacteria are spread primarily by splashing and windblown rain and panicle contact, resulting in the formation of disease foci that are frequently observed in the field [2, 54, 55].

High temperatures in combinations with high humidity or frequent rain are essential for the development of BPB epidemics. The outbreaks of BPB are usually triggered by conditions of high temperatures in combination with simultaneously high relative humidity during the heading-flowering stages. In the observations of Yokoyama and Okuhara [56], the disease developed when minimum daily temperature was ≥23°C and moderate rainfall (<30 mm/day) occurred during heading. Tsushima et al. [57] found BPB commonly occurred when relative humidity was more than 95% for 24 hours during flowering. Lee et al. [58] reported that the disease did not develop when the minimum daily temperature was less than 22°C and when relative humidity was below 80% during the heading stage. Nandakumar et al. [1] found that the optimum temperature for the growth of *B. glumae* and *B. gladioli* ranged from 35 to 40°C.

The outbreaks of BPB in the Southern USA in the epidemic years appeared to be related to unusual weather conditions. Weather conditions favorable for the development of the disease were high nighttime temperatures and high humidity or frequent rainfall during heading and flowering [10]. For example, in the 2010 epidemic year, abnormally high minimum (night time) temperatures occurred on June 21 through July 10 (**Figure 7**) when ca. 60% of the Texas rice acreage was near or at heading and flowering. During that period, rainfall was frequent and relative humidity was 95% or above most of the time (**Figure 7**). The combination of favorable weather conditions, high nighttime temperatures and high humidity, occurring at the most susceptible stages of rice plants promoted the infection and development of BPB. Similar weather patterns were observed in 1995 when a severe epidemic of BPB took place in Texas. There were many days with high maximum temperatures 35°C or

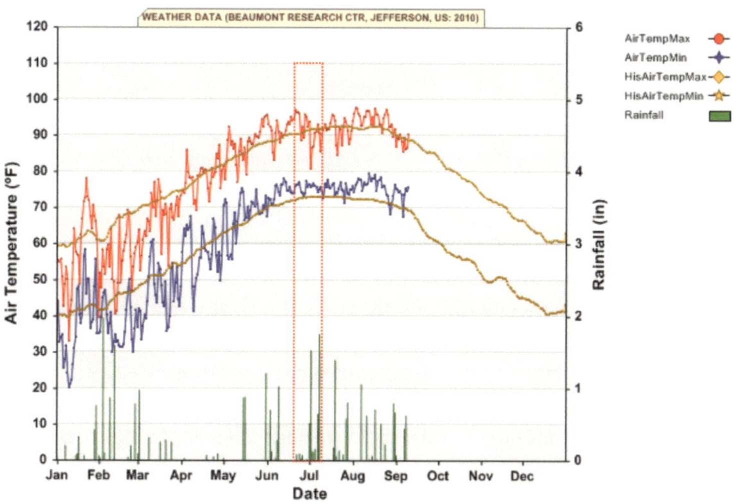

Figure 7.
Air temperatures and rainfalls during the 2010 growing season of rice at the Beaumont Center, Jefferson County, Texas. Note the red-dashed rectangle area showing minimum (night) air temperatures (blue curves) higher above the 65-year historical average (the brown curve) and frequent rainfalls (green bars). The dashed rectangle area represents the period of June 21 through July 10 that coincided with the heading and flowering stages (source: http://beaumont.tamu.edu).

Crop phenology (% heading)	Month	Week	Days				Total precipitation (cm)
			≥35°C	Mean ≥ 24°C	10 am to noon ≥32°C	Precipitation	
—	June	1	0	0	—	0	4.4
7		2	0	3	—	1	2.0
3		3	0	0	0	0	0
6		4	4	1	0	2	8.0
15	July	1	0	5	0	5	4.0
12		2	4	2	0	2	0.7
11		3	4	3	6	2	0.4
10		4	6	5	7	2	4.8
10	August	1	3	1	4	2	2.2
3		2	2	2	6	3	3.5
3		3	4	2	—	3	3.2
3		4	4	1	—	4	2.5

Table 2.
Summary of rice crops and weather data at Beaumont and Eagle Lake, Texas in 1995.

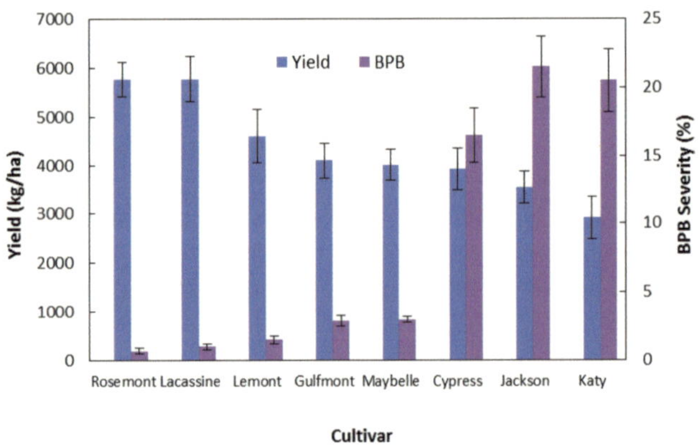

Figure 8.
Yield (left Y-axis) and bacterial panicle blight (BPB) severity (% panicles affected) (right Y-axis) of eight cultivars of rice (X-axis) in naturally infested field at Beaumont, Texas, in 1995 (source: [11]). Error bars are present in columns.

above, day temperatures above 32°C from 10 am to 12 pm (the flowering time), and precipitation from the last week of June through the first week of August (**Table 2**). Heading and flowering occurred on a large percentage of the Texas rice crop during that period. These conditions were associated with severe outbreaks of BPB and significant yield losses in 1995. **Figure 8** shows an example of the severity of this disease in 1995 and its association with yield loss for different rice cultivars, with the disease severity levels ranging from 1 to 22% of panicles affected.

5. Management strategies

Successful disease control generally relies on employing management strategies toward reducing the damage to a manageable and acceptable level. These strategies are exclusion, genetic resistance, chemical control, biological control, and cultural practice. However, for control of the BPB disease at a given geographical area, there are few management options available currently. To effectively manage rice BPB, rice producers must start with the use of pathogen-free seeds as an exclusion measure, plant with partially resistant cultivars, apply with available chemicals or biocontrol agents, and use proper cultural practice. Integrated use of these available management strategies is the key to the effective and sustainable control of the BPB disease.

5.1 Exclusion

Since the BPB disease has been reported in more than 18 countries (**Table 1**) and the disease is not present in all the rice-producing countries and regions, exclusion of the BPB pathogens from a disease-free region is the most effective strategy to prevent BPB of rice. Plant quarantine is an effective measure to achieve this goal. For example, within the USA, the state of California has employed a plant material quarantine procedure to prevent the introduction of the BPB pathogens, other rice pathogens, and weed and insect pests into the state from the southern rice-producing USA. A similar plant quarantine law has been established and enforced in China to prevent the potential importation of the BPB pathogens from foreign countries since 2007 [21].

BPB is seedborne and infected seeds serve as the primary source of inoculum [1, 2, 48]. Therefore, the use of certified seeds that are free of the BPB pathogens is another effective measure to exclude the disease from a disease-free geographic area. Different molecular detection methods including PCR that have been developed to test rice seed lots [19, 48] can aid in this process. In the USA, the use of pathogen-free seeds is recommended to manage the BPB disease. However, using PCR procedure to ensure the BPB pathogens free in certified seed has not been employed. To reduce the BPB disease, it is recommended that farmers should not use the seeds harvested from the fields that are infected with BPB the previous year.

Seed treatment can serve as the last resort to reduce and even eliminate the seedborne BPB pathogen populations and to control subsequent head disease to an accepted level. Rice seeds treated at 65°C of dry heat for 6 days can eradicate the BPB pathogens [26]. Seed treatment with the antibiotic bactericide oxolinic acid (Starner®) has been shown to control the bacterial pathogens in naturally and artificially infected seeds [59]. An antagonistic *Pseudomonas* spp. strain when applied onto seeds was effective to reduce the *B. glumae* populations in seed and suppress seedling rot [60]. Seed treatment with hot water at 60°C for 10 minutes is ineffective for control of the BPB disease although such seed treatment practice is effective to control the rice blast pathogen *M. oryzae* [61].

5.2 Genetic resistance

Considerable research efforts have been conducted globally to develop resistant cultivars as an effective and sustainable strategy for management of BPB of rice. Unfortunately, no single genes or quantitative trait loci (QTLs) for complete resistance to BPB have been found so far [13, 14]. Only several rice cultivars with partial resistance are available for commercial use. In Japan, BPB resistance breeding research efforts started as early as 1975; three partially resistant cultivars were identified through a field screening of nine cultivars and lines [62]. No resistant cultivars and breeding lines were identified in a study of screening 293 cultivars and lines using greenhouse inoculation at the flowering stage in 1983 [63, 64]. From 1985 through 2013, there were nine reported studies that screened a total of 798 cultivars and breeding lines in the field and greenhouse and identified a total of 28 cultivars and lines showing partial resistance to BPB [13, 65–73]. Most recently, Mizobuchi et al.

[74] identified two tropical *japonica* cultivars, Kale and Jaguary, with a high level of resistance and several *indica* cultivars with moderate levels of resistance. These cultivars could serve as good resistance sources to develop BPB-resistant Japanese *temperate japonica* cultivars that can be adapted for use in Japan. Most of rice cultivars commercially available in Japan are susceptible or very susceptible to the BPB disease [74].

In the USA, a collaborative research effort has been established for decades in the southern states of Arkansas, Louisiana, Mississippi, Missouri, and Texas through the Uniform Rice Research Nursery (URRN) to evaluate and develop rice cultivars with high yielding potential and resistance to BPB, sheath blight, rice blast, and other diseases. Annually, more than 200 elite breeding lines and cultivars from the southern states' breeding programs are evaluated in the URRNs inoculated with *B. glumae* at the boot to heading stages. Jupiter, a partially resistant cultivar [75–77] is usually included as a check in these multistate evaluations. Results of multiyear studies demonstrate that no complete resistance cultivars and lines are available and most of the cultivars and lines evaluated are susceptible and very susceptible to BPB [[78], Don Groth, personal communication]. However, some cultivars and lines demonstrated their partial resistance to BPB. For example, Catahoula, Jupiter, Taggart, Rondo, and XL723 (hybrid) were moderately resistant to BPB in the field evaluations conducted in Texas (**Figure 9**). Hybrid cultivars, including XL723, XL753, XL760, CLXL729, CLXL 730, and CLXL745, are relatively more resistant than most of inbred cultivars [4]. The mechanisms associated with BPB resistance in the hybrids are needed to be investigated. In addition, LM-1, a mutant line obtained from gamma radiation treatment of the susceptible cultivar, Lemont, is resistant to BPB [7, 79]. Some resistant breeding lines have been identified in the URRN evaluations in Arkansas [80].

In addition to the host resistance research that has been conducted in Japan and the USA, resistant cultivars and lines have also been reported in other countries. In Brazil, three cultivars were found to be resistant to BPB in the field evaluation [81]. In China, one cultivar, named KaohsiugS.7, was reported to show resistance to the disease when rice plants were inoculated with *B. glumae* at the flowering stage in the field [82].

Host resistance such as rice blast resistance can be broadly classified into complete and partial resistance [83]. The complete resistance is of qualitative character and race specific, which is controlled by major resistance genes (R genes). However, the partial resistance is of quantitative character and non-race specific, which is controlled by several minor genes known as quantitative trait loci (QTLs). Unlike rice blast resistance having both complete and partial resistances, it is apparent

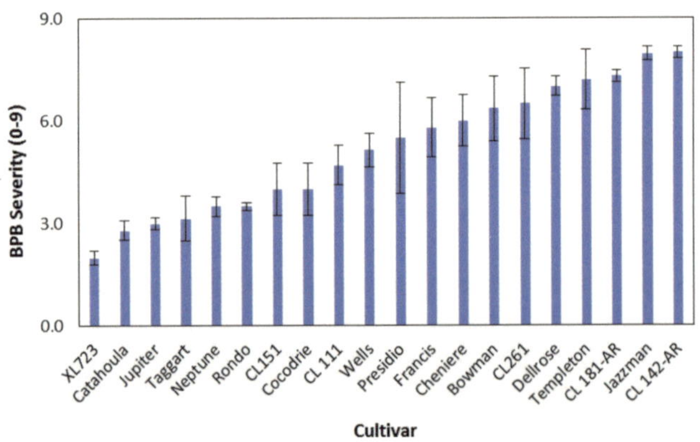

Figure 9.
Mean severities of bacterial panicle blight (BPB) (Y-axis) in 20 rice cultivars (X-axis) over two locations (Beaumont and Eagle Lake) in Texas in 2010. Error bars are present in columns.

that rice BPB resistance has only partial (quantitative) resistance and no complete resistance has been found. Pinson et al. [14] provided the first analysis of QTLs of rice resistance to BPB, using a population of 300 recombinant inbred lines (RILs) derived from a cross between Lemont and TeQing, susceptible and resistant to BPB, respectively. Lemon was an American rice cultivar, while TeQing was a cultivar from China. Twelve QTLs, namely, *qBPB-1-1*, *qBPB-1-2*, *qBPB-1-3*, *qBPB-2-1*, *qBPB-2-2*, *qBPB-3-1*, *qBPB-3-2*, *qBPB-7*, *qBPB-8-1*, *qBPB-8-2*, *qBPB-10*, and *qBPB-11*, were identified on seven chromosomes (chromosomes 1, 2, 3, 7, 8, 10, and 11). Among these QTLs, eight (*qBPB-1-1*, *qBPB-1-2*, *qBPB-2-2*, *qBPB-3-1*, *qBPB-7*, *qBPB-8-1*, *qBPB-10*, and *qBPB-11*) were derived from TeQing and four (*qBPB-1-3*, *qBPB-2-1*, *qBPB-3-2*, and *qBPB-8-2*) from Lemont. After this first report of QTL analysis in the USA, Mizobuchi et al. [73, 84] also identified one QTL, namely, RBG2, on chromosome 1, using a population of 110 backcross inbred lines (BILs) derived from a cross between Kale (resistant to BPB) and Hitomebore (susceptible) in Japan. Kale was a traditional lowland *indica* cultivar that originated from India, while Hitomebore was a modern lowland *temperate japonica* cultivar. In addition, Mizobuchi et al. [85] also have identified the first and only QTL associated with resistance to seedling rot caused by *B. glumae* from a population of 44 chromosome segment substitution lines (CSSLs) derived from a cross between Nona Bohka and Koshihikari, resistant and susceptible to seedling rot, respectively. This QTL, namely, *RBG1*, is located on chromosome 10.

The current research evidence suggests that there is no direct correlation in genetic resistance between seeding rot and grain rot caused by the same bacterium *B. glumae* [64, 73, 85].

5.3 Chemical control

Oxolinic acid (5-ethyl-5,8-dihydro-8-oxo-[1,3]dioxolo[4,5-g]quinoline-7-carboxylic acid, Starner®) is the first chemistry that has been reported to be highly effective for control of the BPB disease in rice. This antibacterial compound, a quinoline derivative, was first introduced in Japan in 1989 for control of rice seeding rot and grain rot [15]. Combined use of oxolinic acid as seed treatment and foliar sprays at heading has been reported to be the best strategy for effective control of both seeding rot and gain rot diseases [17]. When applied at the heading stage, this bactericide is highly effective to inhibit multiplication of *B. glumae* on spikelets and control the BPB disease [15, 51]. In the multiyear field trials conducted in Louisiana is Texas, oxolinic acid, when applied at the boot to heading stages, reduced BPB severity by up to 88% [86–88]. Oxolinic acid has been used three times per season for control of BPB in Japan for more than two decades [89]. Unfortunately, *B. glumae* populations resistant to oxolinic acid have been found in rice in Japan since 1998 [16, 17, 19, 89, 90]. An amino acid substitution at position 83 in GyrA (GyrA83) is responsible for the development of oxolinic acid resistance in the *B. glumae* populations [90]. It has been found that the bacterial populations resistant to oxolinic acid are also cross-resistant to other quinoline derivatives [16]. A specific PCR method has been developed to detect the oxolinic acid-resistant populations of *B. glumae* [19]. The occurrence of oxolinic acid resistance might limit its increasing use and new registrations for management of BPB in rice. Oxolinic acid is not registered for use in rice in the USA and many other countries.

Copper and copper-containing bactericides have also been reported to be effective for control of BPB in rice [86, 91–93]. These bacterial products include Kocide® 2000 (53.8% copper hydroxide), Kocide® 3000 (46.1% copper hydroxide), Previsto® (5% copper hydroxide), Badge® SC (15.4% copper hydroxide plus 16.8% copper oxychloride), Badge® X_2 (21.5% copper hydroxide plus 23.8% copper oxychloride), and Top Cop® (8.4% tric basic copper sulfate). In the field trials of Louisiana, a single application of Kocide® 2000 or Top Cop® at the boot stage reduced the BPB severity as much as 75%, and grain yield and milling quality were improved [86]. In our multiyear field trials conducted in Texas, single applications of Kocide® 3000, Badge® SC, Badge® X_2, or Previsto® at the heading stage significantly reduced BPB severity, with the reductions ranging from 42 to 96% [91–93]. However, except Previsto® with a relatively lower level of copper-active ingredient, all other copper products produced varying degrees of phytotoxicity on sprayed leaves and panicles and under certain environmental conditions reduced yields [86, 91–93]. These copper products have been registered as bactericides and fungicides for control of various bacterial and fungal diseases in citrus, tree crops, vegetables, vines, and field crop (soybeans, wheat, oats, and barley) in the USA. Probably due to their potential phytotoxicity and yield reduction, all these copper products have not been registered for management of the BPB disease on rice in the USA.

In addition to oxolinic acid and copper-based bactericides, other bactericides such as kasugamycin, probenazole, and pyroquilon are used for management of rice seedling rot and grain rot in Japan [16] and Honduras (Lex Ceamer, personal communication).

5.4 Biological control

Several studies have been conducted to develop biological control methods as a strategy for management of BPB of rice. In Japan, Tsushima and Torigoe [94] conducted the first research on the use of bacterial antagonists for control of BPB under field conditions. An antagonistic *Pseudomonas* sp. strain was found to be effective to suppress seedling rot when pretreated onto rice seeds prior to planting [60]. Furuya et al. [95] also found that rice seedling rot was reduced following seed treatment with avirulent strains of *B. glumae*. Miyagawa and Takaya [96] found that an avirulent strain of *B. gladioli* when applied onto rice panicles was very effective to reduce BPB severity. In the USA, five *Bacillus amyloliquefaciens* strains were found to be antagonistic against *B. glumae* in vitro and reduce BPB severity when applied at the heading stage in the field trials conducted in Louisiana [97]. When applied at the flowering stage, two strains of *Bacillus* sp., with antibacterial activities toward *B. glumae*, were demonstrated to reduce BPB severity by as much as 50% and increase grain yield by more than 11% in the field trials conducted in Texas [87, 88]. In a separate BPB-spread field trial study, one of the strains also showed its ability to significantly limit the spatial spread of BPB from a focal point of inoculum [55].

In addition to bacterial biocontrol agents, bacteriophages (also known as phages) have been demonstrated to be effective for management of rice seedling rot in Japan. Adachi et al. [98] found that two bacteriophages were able to lyse *B. glumae* and were highly effective to control seeding rot when rice seeds were pretreated with them. One of the bacteriophages evaluated was even more effective in reducing seeding rot than the bactericide ipconazole/copper (II) hydroxide.

5.5 Cultural practice

Few studies have been conducted to understand and develop cultural practices that could reduce the incidence and severity of BPB in rice. High levels of nitrogen fertility tend to increase the susceptibility of rice plants to the BPB disease. Avoiding excessive nitrogen rates can help reduce the damage caused by BPB. In an Arkansas study evaluating the effects of nitrogen on BPB severity, it was demonstrated that the severity of BPB at the high nitrogen rate (247 kg/ha) was 1.6 times higher than at the low rate (168 kg/ha) applied during a cropping season [99]. Under the Southern

US rice production systems, early planting or use of early maturing rice cultivars to avoid the hottest times of the growing season is another effective approach to reduce the damage caused by the disease. In addition, avoiding excessive seeding rates is also helpful in reducing the incidence and severity of the disease.

6. Conclusion and prospects

BPB has been reported in more than 18 countries and has become a global rice disease. Currently, BPB is one of the major diseases in rice in many countries, including Japan, the USA, and Latin America. The disease is highly destructive, which can cause almost complete losses in yield and milling quality under the most favorable conditions. The outbreaks of BPB are triggered by conditions of high temperatures. With predicted global warming, the disease is likely to be more prevalent on a global scale and to cause more damage in epidemic regions in the future [20, 74]. The global land and ocean surface temperature has been increased by as much as 0.85°C over the period of 1880–2012 based on the 2014 IPCC report [100]. Under the 1°C warming scenario, it is estimated that the increased damage caused by this disease in the Southern USA would result in a $103 million USD annual decrease in consumer surplus and a loss of rice production equivalent to feeding 1.9 million people (Aaron Shew, personal communication).

Effective management of this bacterial disease is challenging. Unlike most of other rice diseases, The BPB disease often develops after the heading stage, and typically no symptoms and signs can be observed before heading. Therefore, no scouting methods are currently available to detect and predict the development of the disease. No standardized seed treatment methods have been developed and commercialized specifically to eradicate or reduce the pathogen populations in rice seeds. No chemical control agents are labeled for management of the BPB disease in most countries, including the USA. The efficacy and increasing use of oxolinic acid have been affected by the development of oxolinic acid resistance in the populations of *B. glumae* in Japan and other countries. No commercially available biocontrol agents have been developed. Most of commercially available rice cultivars are susceptible or very susceptible to BPB.

Therefore, effective and sustainable control of the BPB disease largely depends on integrated use of available management options. Plant quarantine is the first defense to exclude the BPB pathogens from disease-free countries and regions. The use of pathogen-free seed or certified seed is another effective measure to control this disease. Planting with cultivars having a resistant level as high as possible is always an effective recommendation to reduce the damage caused by the disease. A limited number of rice cultivars, including hybrids, with partial resistance to BPB are available for commercial use in many countries. Since no source of complete resistance has been discovered so far, more research is needed to look for new sources of resistance through screening a greater number of germplasm lines, including those from other countries and the wild species of *Oryza*. Continued studies are needed to further characterize, fine map, or even clone the QTLs associated with BPB resistance that have been identified. More investigations are desired to understand the genetic control of BPB resistance in available resistant rice cultivars and lines, especially hybrids. These studies may lead to the development of molecular makers linked to BPB resistance that can help breeders facilitate the selection of BPB resistance in early breeding generations with more confidence. Recent advances in rice genomics and newly developed genome editing tools like CRISPR may provide new and powerful tools to better understand the mechanisms associated with BPB resistance and develop new rice cultivars with a higher level of resistance to BPB in the future. Developing and use of resistant cultivars is the best strategy to minimize the damage caused by BPB and maximize rice production in the long term.

Author details

Xin-Gen Zhou
Texas A&M AgriLife Research and Extension Center, Texas A&M University System, Beaumont, Texas, USA

*Address all correspondence to: xzhou@aesrg.tamu.edu

References

[1] Nandakumar R, Shahjahan AKM, Yuan XL, Dickstein ER, Groth DE, Clark CA, et al. *Burkholderia glumae* and *B. gladioli* cause bacterial panicle blight in rice in the southern United States. Plant Disease. 2009;**93**:896-905

[2] Tsushima S. Epidemiology of bacterial grain rots of rice caused by *Pseudomonas glumae*. Japan Agricultural Research Quarterly. 1996;**30**:85-89

[3] Groth DE, Hollier C, Rush C: Disease management. In: Saichuk J, editor. Louisiana Rice Production Handbook. Crowley, Louisiana: The LSU AgCenterPub; 2321 (3M); 2009. pp. 46-92

[4] Zhou XG, Jo Y-K. Disease management. In: Way MO, McCauley GM, Zhou XG, Wilson LT, Brandy M, editors. 2014 Texas Rice Production Guidelines. Beaumont, Texas: Texas A&M AgriLife Research and Texas A&M AgriLife Extension; B-6131; 2014. pp. 44-57. Available from: https://beaumont.tamu.edu/eLibrary/Bulletins/2014_Rice_Production_Guidelines.pdf

[5] JPPA. Epidemic and controlling areas in 2013. In: Catalogue of Agricultural Chemicals. Tokyo: Japan Plant Protection Association; 2014. pp. 572-574

[6] JPPA. Epidemic and controlling areas in 2015. In: Catalogue of Agricultural Chemicals. Tokyo: Japan Plant Protection Association; 2016. pp. 562-564

[7] Groth DE, Linscombe SD. Rice resistance to sheath blight, bacterial panicle blight, and other diseases. The US Patent Application Publication. US2004/0168232; 2004. 1-5

[8] Shahjahan AKM, Rush MC, Groth DE, Clark CA. Panicle blight. Rice. 2000;**15**:26-29

[9] Mulaw T, Wamishe Y, Jia Y. Characterization and in plant detection of bacteria that cause bacterial panicle blight of rice. American Journal of Plant Sciences. 2018;**9**:667-684

[10] Zhou XG, McClung AM, Way MO, Jo Y-K, Tabien RE, Wilson LT. Severe outbreaks of bacterial panicle blight across Texas Rice belt in 2010. Phytopathology. 2011;**101**:S205

[11] Zhou XG, McClung AM, Way MO, Tabien RE, Wilson LT. Severe outbreaks of bacterial panicle blight in Texas. Texas Rice. 2010;**10**:3-8. Available from: https://beaumont.tamu.edu/eLibrary/Newsletter/2010_September_Newsletter.pdf

[12] Zhou XG, Liu G. Quantification of yield losses caused by bacterial panicle blight in rice. Phytopathology. 2012;**102**:S4.144

[13] Mizobuchi R, Fukuoka S, Tsushima S, Yano M, Sato H. QTLs for resistance to major rice diseases exacerbated by global warming: Brown spot, bacterial seedling rot, and bacterial grain rot. Rice. 2016;**9**:23

[14] Pinson SRM, Shahjahan AKM, Rush MC, Groth DE. Bacterial panicle blight resistance QTLs in rice and their association with other disease resistance loci and heading date. Crop Science. 2010;**50**:1287-1297

[15] Hikichi Y, Noda C, Shimizu K. Oxolinic acid. Japanese Journal of Infectious Diseases. 1989;**55**:21-23

[16] Hikichi Y, Egami H, Oguri Y, Okuno T. Fitness for survival of *Burkholderia glumae* resistant to oxolinic acid in rice plants. Annals of the Phytopathological Society of Japan. 1998;**64**:147-152

[17] Hikichi Y, Tsujiguchi K, Maeda Y, Okuno T. Development of increased oxolinic acid-resistance in *Burkholderia glumae*. Journal of General Plant Pathology. 2001;**67**:58-62

[18] Maeda Y, Horita M, Shinohara H, Kiba A, Ohnish K, Tsuchima S, et al. Analysis of sources of oxolinic acid-resistant field strains of *Burkholderia glumae* based on rep-PCR analysis and nucleotide sequences of *gyrB* and *rpoD*. Journal of General Plant Pathology. 2007;**73**:46-52

[19] Maeda Y, Kiba A, Ohnish K, Hikichi Y. New method to detect oxolinic acid-resistant *Burkholderia glumae* infecting rice seeds using a mismatch amplification mutation assay polymerase chain reaction. Journal of General Plant Pathology. 2004;**70**:215-217

[20] Ham JH, Melanson RA, Rush MC. *Burkholderia glumae*: Next major pathogen of rice? Molecular Plant Pathology. 2011;**12**:329-339

[21] Cui Z-G, Zhu B, Xie G-L, Li B, Huang S-W. Research status and prospect of *Burkholderia glumae*, the pathogen causing bacterial panicle blight. Rice Science. 2016;**23**:111-118

[22] Goto K, Ohata K. New bacterial disease of rice (brown stripe and grain rot). Annals of the Phytopathological Society of Japan. 1956;**21**:46-44

[23] USDA/FAS: Grain: World Market and Trade. USDA/FAS/Office of Global Analysis; September 2018. 39

[24] Kurita T, Tabei H. On the pathogenic bacterium of bacterial grain rot of rice. Annals of the Phytopathological Society of Japan. 1967;**33**:111

[25] Chien CC, Chang YC, Liao YM. Bacterial grain rot of rice: A new disease in Taiwan. Journal of Agricultural Research of China. 1983:360-366

[26] Zeigler RS, Alvarez E. Grain discoloration of rice caused by *Pseudomonas glumae* in Latin America. Plant Disease. 1989;**73**:368

[27] Trung HM, Van NV, Vien NV, Lam DT, Lien M. Occurrence of rice grain rot disease in Vietnam. International Rice Research Notes. 1993;**18**:30

[28] Ura H, Iiyama K, Furuya N, Matsuyama N. Properties of the bacteria isolated from rice plants with leaf-sheath browning and grain rot. Annals of the Phytopathological Society of Japan. 1996;**62**:640

[29] Ura H, Furuya N, Iiyama K, Hidaka M, Tsuchiya K, Matsuyama N. *Burkholderia gladioli* associated with symptoms of bacterial grain rot and leaf-sheath browning of rice plants. Journal of General Plant Pathology. 2006;**72**:98-103

[30] Cottyn B, Cerez MT, VanOutryve MF, Barroga J, Swings J, Mew TW. Bacterial diseases of rice. 1. Pathogenic bacteria associated with sheath rot complex and grain discoloration of rice in the Philippines. Plant Disease. 1996;**80**:429-437

[31] Cottyn B, VanOutryve MF, Cerez MT, DeCleene M, Swings J, Mew TW. Bacterial diseases of rice. 2. Characterization of pathogenic bacteria associated with sheath rot complex and grain discoloration of rice in the Philippines. Plant Disease. 1996;**80**:438-445

[32] Cottyn B, Regalado E, Lanoot B, De Cleene M, Mew TW, Swings J. Bacterial populations associated with rice seed in the tropical environment. Phytopathology. 2001;**91**:282-292

[33] Yuan XL. Identification of Bacterial Pathogens Causing a Blight (Master thesis). Louisiana State University; 2005

[34] Jeong Y, Kim J, Kim S, Kang Y, Nagamatsu T, Hwang I. Toxoflavin produced by *Burkholderia glumae* causing rice grain rot is responsible for inducing bacterial wilt in many field crops. Plant Disease. 2003;**87**:890-895

[35] Luo JY, Xie GL, Li BQ, Xu LH. First report of *Burkholderia glumae* isolated

from symptomless rice seeds in China. Plant Disease. 2007;**91**:1363

[36] Nandakumar R, Rush MC, Correa F. Association of *Burkholderia glumae* and *B. gladioli* with panicle blight symptoms on rice in Panama. Plant Disease. 2007;**91**:767

[37] CIAT: Annual Report of SBA-4: Rice. CIAT; March 2008. 30

[38] Lu SE, Allen TW. Causal agents of bacterial panicle blight of rice and evaluation of disease resistance of rice cultivars to the disease in Mississippi. In: 34th Proceedings of Rice Technical Working Group (RTWG). Vol. 2012. Arkansas, USA: Hot Spring; 2012. p. 76

[39] Quesada-González A, García-Santamaría F. *Burkholderia glumae* in the rice crop in Costa Rica. Agronomía Mesoamericana. 2014;**25**:371-381

[40] Riera-Ruiz C, Vargas J, Cevallos-Cevallos JM, Ratti M, Peralta EL. First report of *Burkholderia glumae* causing bacterial panicle blight on rice in Ecuador. Plant Disease. 2014;**98**:988

[41] Zhou XG. First report of bacterial panicle blight of rice caused by *Burkholderia glumae* in South Africa. Plant Disease. 2014;**98**:566

[42] Mondal KK, Mani C, Verma G. Emergence of bacterial panicle blight caused by *Burkholderia glumae* in North India. Plant Disease. 2015;**99**:1268

[43] Mirghasempour SA, Huang S, Xie GL. First report of *Burkholderia gladioli* causing rice panicle blight and grain discoloration in China. Plant Disease. 2018;**102**:263

[44] Shahjahan AKM, Rush MC, Clark CA, Groth DE. Bacterial sheath rot and panicle blight of rice in Louisiana. In: Proceedings of 27th Rice Technical Working Group (RTWG). Texas, USA: College Station; 1998, 1998. pp. 131-132

[45] Iiyama K, Furuya N, Takanami Y, Matsuyama N. A role of phytotoxin in virulence of *Pseudomonas glumae* Kurita et Tabei. Annals of the Phytopathological Society of Japan. 1995;**61**:470-476

[46] Matsuda I, Sato Z. Relations between pathogenicity and pigment productivity in the casual agent of bacterial grain rot of rice. Annals of the Phytopathological Society of Japan. 1988;**54**:378

[47] Natalie AH, Castro BA, McDonald EM, Pellerano MA, Ochoa R. The panicle rice mite, Steneotarsonemus spinki smiley, a re-discovered pest of rice in the United States. Crop Protection. 2009;**28**:547-560

[48] Sayler RJ, Cartwright RD, Yang Y. Genetic characterization and real-time PCR detection of *Burkholderia glumae*, a newly emerging bacterial pathogen of rice in the United States. Plant Disease. 2006;**90**:603-610

[49] Compant S, Nowak J, Coenye T, Clement C, Barka EA. Diversity and occurrence of *Burkholderia* spp. in the natural environment. FEMS Microbiology Reviews. 2008;**32**:607-626

[50] Nandakumar R, Bollich PA, Shahjahan AKM, Groth DE, Rush MC. Evidence for the soilborne nature of the rice sheath rot and panicle blight pathogen, *Burkholderia gladioli* can. Journal of Plant Pathology. 2008;**30**:148-154

[51] Hikichi Y. Relationship between population dynamics of *Pseudomonas glumae* on rice plants and disease severity of bacterial grain rot of rice. Journal of Pesticide Science. 1993;**18**:319-324

[52] Hikichi Y, Okuna T, Furusawa I. Susceptibility of rice spikelets to infection with *Pseudomonas glumae* and its population dynamics. Journal of Pesticide Science. 1999;**19**:11-17

[53] Li L, Wang L, Liu L-M, Hou Y-X, Li Q-Q, Huang S-W. Infection process of *Burkholderia glumae* before booting stage of rice. Journal of Phytopathology. 2016;**164**:825-832

[54] Tsuchima S, Naito H. Spatial distribution and dissemination of bacterial grain rot of rice caused by *Pseudomonas glumae*. Annals of the Phytopathological Society of Japan. 1991;**57**:180-187

[55] Zhou XG. Effects of cultivar resistance and biocontrol agent on spatial spread of bacterial panicle blight of rice. Phytopathology. 2014, 2014;**104**:S3. 137

[56] Yokoyama T, Okuhara K. Relationship between the occurrence of rice bacterial grain rot and meteorological conditions and cultivars of rice. Proceedings of the Association for Plant Protection of Kyushu. 1987;**33**:4-8

[57] Tsushima S, Naito H, Koitabashi M. Change in panicle susceptibility associated with flowering rate of spikelets in bacterial grain rot of rice caused by *Pseudomonas glumae*. Japanese Journal of Phytopathology. 1995;**61**:109-113

[58] Lee YH, Ko S-J, Cha K-H, Park EW. BGRcast: A disease forecast model to support decision-making for chemical sprays to control bacterial grain rot of rice. Plant Pathology Journal. 2015;**31**:350-362

[59] Wamishe Y, Cartwright R, Belmar S, Kelsey C. Preliminary studies on management of seed-borne inoculum of *Burkholderia glumae* in rice seed. In: 34th Proceedings of Rice Technical Working Group (RTWG). Arkansas, USA: Hot Spring; 2012, 2012. p. 94

[60] Inoue H, Nakaho K, Miyagawa H. Efficacy of the antagonistic bacteria CAB-02 on prevention of bacterial grain rot and seedling blight of rice, applied together with seed disinfectants. Phytopathology. 2001;**90**:S37

[61] Hayasaka T, Ishiguro K, Shibutani K, Namai T. Seed disinfection using hot water immersion to control several seed-borne diseases of rice plants. Japanese Journal of Phytopathology. 2001;**67**:26-32

[62] Goto T, Watanabe B. Varietal resistance to bacterial grain rot of rice, caused by *Pseudomonas glumae* (in Japanese). Proceedings of the Association for Plant Protection of Kyushu. 1975;**21**:141-143

[63] Goto T. Inoculation methods for resistance to bacterial seedling rot and grain rot in *Burkholderia glumae*. (in Japanese). Annals of the Phytopathological Society of Japan. 1982;**48**:373

[64] Goto T. Resistance to *Burkholderia glumae* of lowland rice cultivars and promising lines in Japan. Annals of the Phytopathological Society of Japan. 1983;**49**:410

[65] Mogi S, Tsushima S. Varietal resistance to bacterial grain rot in rice, caused by *pseudomonas glumae*. National Agricultural Research. 1985;**47**:103

[66] Imbe T, Tsushima S, Nishiyama H. Varietal resistance of rice to bacterial grain rot and screening method. Proceedings of the Association for Plant Protection of Kyushu. 1986;**32**:17-19

[67] Yokoyama T, Okuhara K. Relationship between the occurrence of rice bacterial grain rot, and meteorological conditions and cultivars of rice. (in Japanese). Proceedings of the Association for Plant Protection of Kyushu. 1987;**33**:4-8

[68] Takita T, Imbe T, Nishiyama H, Tsushima S. Resistance to rice bacterial grain rot in indica and upland rice.

National Agricultural Research. 1988;**50**:28

[69] Tsushima S, Mogi S, Naito H, Saito H. Existence of *pseudomonas glumae* on the rice seeds and development of the simple method for detecting *P. glumae* from the rice seeds. Bulletin of the Kyushu National Agricultural Experiment Station. 1989;**25**:261-270

[70] Miyagawa H, Kimura T. A test of rice varietal resistance to bacterial grain rot by inoculation on cut-spikes at anthesis. Bulletin of the Chugoku National Agricultural Experiment Station. 1989;**78**:17-21

[71] Wasano K, Okuda S. Evaluation of resistance of rice cultivars to bacterial grain rot by the syringe inoculation method. Breeding Science. 1994;**44**:1-6

[72] Yasunaga T, Wada T, Oosata KF, Hamachi Y. Varietal differences in occurrence of bacterial grain rot in rice cultivars with high palatability. Crop Science Society of Japan. 2002;**68**:12-14

[73] Mizobuchi R, Sato H, Fukuoka S, Tanabata T, Tsushima S, Imbe T, et al. Mapping a quantitative trait locus for resistance to bacterial grain rot in rice. Rice. 2013;**6**:13

[74] Mizobuchi R, Fukuoka S, Tsuiki C, Tsushima S, Sato H. Evaluation of major Japanese rice cultivars for resistance to bacterial grain rot caused by *Burkholderia glumae* and identification of standard cultivars. Breeding Science. 2018;**68**:413-419

[75] Sha X, Linscombe SD, Groth DE, Bond JA, White LM, Chu QR, et al. Registration of Jupiter' rice. Crop Science. 2006;**46**:1811-1812

[76] Nandakumar R, Bollich PA, Groth DE, Rush MC. Confirmation of the partial resistance of Jupiter rice to bacterial panicle blight caused by *Burkholderia glumae* through reduced disease and yield loss in inoculated field tests. Phytopathology. 2007;**97**:S82

[77] Nandakumar Rush MC. Analysis of gene expression in Jupiter rice showing partial resistance to rice panicle blight caused by *Burkholderia glumae*. Phytopathology. 2008;**98**:S112

[78] Zhou XG, Tabien RE, McClung AM. Field disease reaction of rice cultivars and elite lines in Texas. Phytopathology. 2011;**101**:S204

[79] Groth DE, Linscombe SD, Sha X. Registration of two disease resistant germplasm lines of rice. Journal of Plant Registrations. 2007;**1**:63-64

[80] Wamishe Y, Jia Y, Rasheed M, Kelsey C, Belmar S, Gebremariam T. Resistance the ultimate solution for bacterial panicle blight disease of rice. Phytopathology. 2014;**104**:S124

[81] Prabhu AS, Bedendo IP. Glume blight of rice in Brazil: Etiology, varietal reaction and loss estimates. Tropical Pest Management. 1988;**34**:85-88

[82] Chien CC, Chang YC. The susceptibility of rice plants at different growth stages and 21 commercial rice varieties to *Pseudomonas glumae*. Journal of Agricultural Research of China. 1987;**36**:302-310

[83] Wang GL, Mackill DJ, Bonman JM, McCouch SR, Champoux MC, Nelson RJ. RFLP mapping of genes conferring complete and partial resistance to blast in a durably resistant rice cultivar. Genetics. 1994;**136**:1421-1434

[84] Mizobuchi R, Sato H, Fukuoka S, Tsushima S, Imbe T, Yano M. Fine mapping of *RBG2*, a quantitative trait locus for resistance to *Burkholderia glumae*, on rice chromosome 1. Molecular Breeding. 2015;**35**:15

[85] Mizobuchi R, Sato H, Fukuoka S, Tsushima S, Imbe T, Yano M.

Identification of *qRBS1*, a QTL involved in resistance to bacterial seedling rot in rice. Theoretical and Applied Genetics. 2013;**126**:2417-2425

[86] Groth DE, Shahjahan AKM, Rush MC. Control of bacterial panicle blight of rice with foliar applications of bactericides. Phytopathology. 2001;**91**:S33

[87] Zhou XG, Liu G, Kloepper JW, Reddy MS, Kumar KVK, Zhang S. Management of bacterial panicle blight of rice with beneficial *Bacillus* strains. Phytopathology. 2011;**101**:S270

[88] Zhou XG, Liu G, Kloepper JW, Reddy MS, Kumar KVK, Zhang S. Evaluation of beneficial *Bacillus* strains for suppression of bacterial panicle blight in rice, 2010. Plant Disease Management Reports. 2011;**5**:FC079

[89] Maeda Y, Kiba A, Ohnishi K, Hikichi Y. Amino acid substitutions in Gyr a of *Burkholderia glumae* are implicated in not only oxolinic acid resistance but also fitness on rice plants. Applied and Environmental Microbiology. 2007;**73**:1114-1119

[90] Maeda Y, Kiba A, Ohnishi K, Hikichi Y. Implications of amino acid substitutions in GyrA at position 83 in terms of oxolinic acid resistance in field strains of *Burkholderia glumae*, a causal agent of bacterial seedling rot and grain rot of rice. Applied and Environmental Microbiology. 2004;**70**:5613-5620

[91] Zhou XG, Liu G. Field evaluation of bactericides for control of bacterial panicle blight in rice, 2013. Plant Disease Management Reports. 2014;**8**:FC232

[92] Zhou XG, Liu G. Field evaluation of badge SC and badge X2 for control of bacterial panicle blight in rice, 2013. Plant Disease Management Reports. 2014;**8**:FC231

[93] Zhou XG, Liu G, Zhou L. Evaluation of bactericides for control of bacterial panicle blight of rice in Texas, 2016. Plant Disease Management Reports. 2017;**11**:FC118

[94] Tsushima S, Torigoe H. Suppression of bacterial grain rot of rice by antagonistic bacteria (in Japanese). Plant Protection. 1991;**3**:91-95

[95] Furuya N, Okamoto T, Kori Y, Matsuyama N, Wakimoto S. Control of bacterial seedling rot of rice by avirulent strains of *Pseudomonas glumae*. Annals of the Phytopathological Society of Japan. 1991;**57**:371-376

[96] Miyagawa H, Takaya S. Biological control of bacterial grain rot of rice by avirulent strain of *Burkholderia gladioli*. Bulletin of the Chugoku National Agricultural Experiment Station. 2000;**21**:1-21

[97] Shrestha BK, Karki HS, Groth DE, Jungkhun N, Ham JH. Biological control activities of rice-associated *Bacillus* sp. strains against sheath blight and bacterial panicle blight of rice. PLoS One. 2016;**11**(1):e0146764. DOI: 10.1371/journal.pone.0146764

[98] Adachi N, Tsukamoto S, Inoue S, Azegami K. Control of bacterial seedling rot and seedling blight of rice by bacteriophage. Plant Disease. 2012;**96**:1033-1036

[99] Wamishe YA, Belmar SB, Kelsey CD, Gebremariam TA, McCarty DL. Effects of excessive nitrogen fertilizer on rice diseases with emphasis to bacterial panicle blight. In: 35th Proceedings of Rice Technical Working Group (RTWG). New Orleans, Louisiana, USA; 2014. pp. 86-87

[100] IPCC: Climate change 2014, synthesis report, summary for policymakers. 2014. Available from: http://www.ipcc.ch/report/ar5/syr/

Section 2

Emerging New Threats to Rice Productivity and Quality

Chapter 7

Rice False Smut: An Increasing Threat to Grain Yield and Quality

Wen-Ming Wang, Jing Fan and John Martin Jerome Jeyakumar

Abstract

Rice false smut (RFS) is the most important grain disease in rice production worldwide. Its epidemics not only lead to yield loss but also reduce grain quality because of multiple mycotoxins generated by the causative pathogen, *Villosiclava virens* (anamorph: *Ustilaginoidea virens*). The pathogen infects developing spikelets and specifically converts individual grain into a RFS ball that is established from mycelia covered with powdery chlamydospores, sometimes generating sclerotia. RFS balls seem to be randomly formed in some grains on a panicle of a plant in the paddy field. However, epidemics differ largely among varieties, fields, and seasons. This chapter introduces current understanding on the disease, mycotoxins, the biology of the pathogen, pathogenesis of RFS, rice resistance, the disease cycle, the disease control, and assay.

Keywords: basal defense, biotroph, effector, epiphytic growth, grain filling gene, mycotoxin, sclerotium

1. Introduction

Rice production plays a crucial role in our food security. Rice security is not only an economic issue but also an important parameter to determine social and political stability [1]. Thus, rice research has to be geared up to develop strategies for alleviating losses due to pests and diseases. In the past decades, a number of minor diseases have attained the status of major importance in rice. One such disease is the rice false smut (RFS) disease that is a threat to yield and grain quality.

RFS was previously recorded as a minor disease of rice and considered as a symbol of good harvest in old times. In recent years, increasing occurrence of RFS has been reported in most major rice growing regions throughout the world, such as China, India, and USA [2–5]. The emergence of this disease is believed to be partially due to wide application of hybrid rice varieties, which are mostly susceptible to the RFS. The causative agent of RFS is an ascomycete fungal pathogen *Villosiclava virens* (anamorph: *Ustilaginoidea virens* [Cooke] Takahashi) [6], which specifically infects rice flowers and transforms the latter into RFS balls [3]. RFS balls are small at first growing slowly and enclosing the floral parts. The early balls were found to be slightly flattened and smooth and were covered by a thin membrane. As the pathogen growth intensifies, the RFS ball bursts with chlamydospores and becomes orange then later yellowish-green or greenish-black (**Figure 1A–C**). The RFS balls generate sclerotia (**Figure 1D**) when the temperature difference between day and night is large in autumn [3]. RFS ball is the only visible symptom of RFS disease.

IntechOpen

Figure 1.
Disease symptom of rice false smut: (A)–(C) white, yellow, and dark green false smut balls at early, middle, and late stages, respectively; (D) sclerotia (white arrows) are formed in false smut balls at the late stage; (E) field view of rice false smut disease; and (F) harvested rice grains are contaminated with rice false smut balls. Inset shows that rice grains are covered by chlamydospores from false smut balls.

The disease induces considerable losses both in yield and quality [7, 8], due to the occurrence of RFS balls and increased sterility of kernels adjacent to the balls [9]. Moreover, RFS balls produce two types of mycotoxins, i.e., ustiloxin and ustilaginoidin, which are poisonous to both humans and animals and impose significant health hazards by contaminating rice grains and straws [10–12]. For example, ustiloxin A causes kidney and liver damage in mice, due to its inhibition activity on microtubule assembly and skeleton formation of the eukaryotic cells [11, 13].

RFS balls seem to be randomly formed in some grains of rice panicles in the paddy field and are inevitably collected during harvest (**Figure 1E, F**). The disease spread varies within a field or between fields and is considered to be more severe in the proximity of drainage channels [14]. Epidemics of RFS disease tend to occur when rice booting and heading stages meet with rainfall periods. However, epidemics differ largely among varieties, fields, and seasons. This chapter describes our current knowledge on the mycotoxins, biology of the pathogen, pathogenesis of RFS, rice resistance, disease cycle, disease control, and disease assay.

2. Mycotoxins

V. virens produces two kinds of mycotoxins, ustiloxins and ustilaginoidins. Ustiloxins are water-soluble cyclic peptides, each including a 13-membered ring with an ether linkage [10]. The structures of six ustiloxins, including A–D, F, and G, have been identified so far [10, 15, 16]. A gene cluster has been suggested to be responsible for the ribosomal biosynthesis of ustiloxins [17]. Ustiloxins A and B are among the most abundant ustiloxins in RFS ball and are mainly contained in the middle layer of mycelia and immature chlamydospores at early maturity stage [18]. Recently, rapid qualitative or quantitative detection methods, such as monoclonal antibody-based enzyme-linked immunosorbent assay and colloidal gold-based lateral flow immunoassay, have been established for detecting ustiloxins A and B in rice and feed samples [19, 20]. Ustilaginoidins are bis-naphtho-γ-pyrones and can be easily dissolved in organic solvent. Twenty-five ustilaginoidins (A-W) have been isolated from *V. virens* [21–25]. Ustilaginoidins can be isolated from RFS balls and solid rice media culturing *V. virens*. Several tested ustilaginoidins are mainly distributed in the layers embracing chlamydospores of yellowish-green or dark-green RFS balls [26].

Ustiloxins could inhibit polymerization of microtubule proteins and cause abnormal mitosis resembling, which would result in acute necrosis of renal tubular cells and hepatocytes in mice [10]. Ustiloxins also show phytotoxicity, inhibiting elongation of radicle and germ and inducing swelling of seedling root in rice [10, 16]. Cytotoxic activities of ustiloxins have been demonstrated on human tumor cell lines, such as A375, A549, BGC-823, HCT116, and HepG2 [10, 16]. Similar phytotoxic and cytotoxic activities have been detected for ustilaginoidins [25, 27, 28]. In addition, ustilaginoidins show antibacterial activities against several human or plant pathogens, such as *Agrobacterium tumefaciens*, *Bacillus subtilis*, *Pseudomonas lachrymans*, *Ralstonia solanacearum*, *Staphylococcus haemolyticus*, and *Xanthomonas vesicatoria* [24]. Due to their anti-tumor and/or antibacterial abilities, ustiloxins and ustilaginoidins may be used as potential clinical medications. As a result, chemical synthesis of ustiloxins and their analogs has been carried out [29, 30].

3. Biology of the pathogen

The RFS pathogen belongs to the kingdom: Fungi, phylum: Ascomycota, class: Ascomycetes, subclass: Sordariomycetes, order: Hypocreales, family: Clavicipitaceae, genus: *Villosiclava,* and species: *virens* and its anamorphic stage is *Ustilaginoidea virens* [6]. The colony growth of *V. virens* in culture medium PSA (potato-sucrose-agar) is very slow, with a growth rate of approximately 20 mm in diameter per week [31]. *V. virens* produces pigments during culture in PSA and is prone to generate small colonies and plenty of conidia in PSB (potato-sucrose-broth) (**Figure 2A–D**). The conidia are elliptical with diameters ranging from 3 to 5 μm (**Figure 2E**). Upon maturation or under unfavorable conditions, conidia may develop to rounded chlamydospores with prominent spines on the surface (**Figure 2F**) [32–34]. One or two sclerotia, which are the sexual structure of *V. virens*, can be formed in a RFS ball (**Figure 1D**). Sclerotia are horseshoe-shaped and the length ranged from 2 to 20 mm (**Figure 2G**). After several months of dormancy, sclerotia could germinate and produce fruiting bodies with stromata (**Figure 2F**), which ultimately generates ascospores with length reaching 50 μm and width 1 μm [35].

Numerous efforts have been undertaken to optimize the culture media and culturing conditions for *V. virens*. PSA and PDA (potato-dextrose-agar) are suitable for culturing *V. virens* in solid media [31]. Moreover, stachyose is a preferential carbon source for *V. virens* and could significantly promote hyphal growth and conidia germination of *V. virens*, much better than other sugars, such as sucrose, glucose, and starch [36]. Stachyose can be also applied in optimization of culture medium for other filamentous fungi. Ammonium chloride, ammonium sulfate, and ammonium nitrate are the suitable nitrogen sources for *V. virens* growth [31]. The optimal growth of *V. virens* can be achieved at 28°C and pH 6–7 [37].

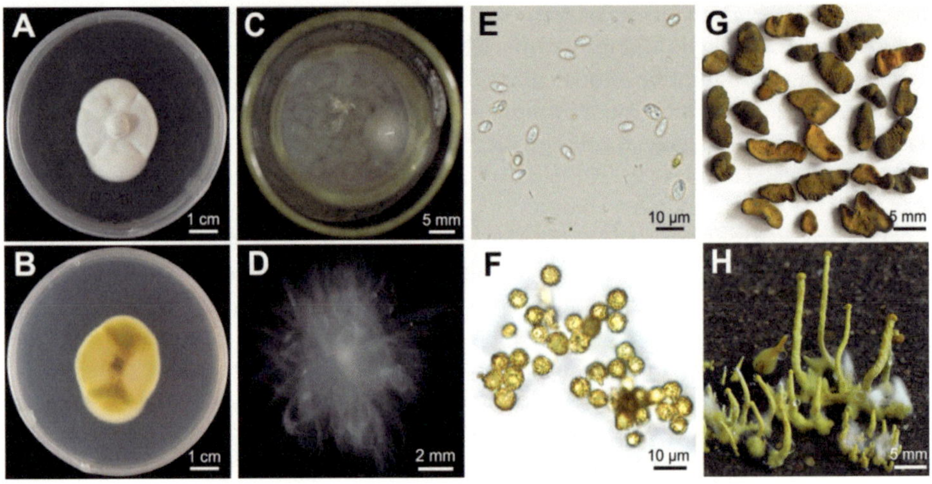

Figure 2.
Morphology of the rice false smut pathogen. Colony of Ustilaginoidea virens (anamorph) in PSA medium from top view (A) and back view (B). (C) Colonies of U. virens in PSB. (D) A single colony of U. virens in PSB. (E) Conidia of U. virens in PSB. (F) Chlamydospores of U. virens from false smut balls. (G) Sclerotia of Villosiclava virens (telemorph). (H) Sclerotia of V. virens germinate to produce fruit bodies. Images (G) and (H) are courtesy of Prof. Dongwei Hu from Zhejiang University, China.

4. Pathogenesis of RFS

4.1 Infection process of *V. virens*

The RFS pathogen *V. virens* specifically attacks rice flowers to form RFS balls, causing economically important disease. The infection process of *V. virens* in rice flower has been identified cytologically as follows: at late booting stage of rice, spores of *V. virens* come into contact with developing spikelets and germinate on their surface, or epiphytically grown hyphae reach the surface of developing spikelets. The hyphae could not penetrate the spikelet but extend into the inner space of a spikelet via the gap between the lemma and the palea [38]. After entering, the pathogen primarily infects the stamen filaments intercellularly [39], probably due to loose alignment of cells and flexible cell walls [40]. Lodicules and stigma could also be attacked, although to a lesser extent [39, 41]. However, no infection structures, such as appressorium and haustorium, can be detected during infection. Along with time, mycelia grow to enclose all the floral organs, then protrude out of the spikelet, and ultimately form a ball-shape colony covered with chlamydospores. At late stage of infection or in a RFS ball, stamen filaments are replaced by mycelia, but the ovary and lodicules remain intact, suggesting that they may contribute to the formation of RFS ball [42]. In addition, the hyphae of *V. virens* could not extend into pedicles and stems connecting the spikelets, and no anatomic changes are detected in pedicles [39].

Although RFS disease symptoms are observed at rice grains due to pathogen infections of spikelets, *V. virens* also grows on other rice organs without obvious symptoms. At the germination stage of rice seed, chlamydospores of *V. virens* could germinate on coleoptiles and the hyphae are able to extend intercellularly between epidermal cells [43]. At seedling stage, chlamydospores could also germinate on the surface of roots and grow in the intercellular space of root epidermal cells [44, 45]. More recently, a detailed observation on *V. virens*-infected rice roots indicates that the cellulose microfibrils of epidermal cell wall are very loose, similar to those of stamen filaments, and thus are prone to be infected [46]. However, the hyphae of *V. virens* are stopped by sclerenchyma layer from entering into endodermis and phloem tissues [46]. Again, no appressorium or haustorium can be detected when *V. virens* infects tender coleoptiles and roots. To date, contribution of the infections of the coleoptiles and roots at the vegetative stage to RFS disease symptoms has not been determined if any, and the evidence of systemic infection of *V. virens* is currently lacking.

4.2 Genetic transformation of *V. virens*

Genetic manipulation is essential to clarify the pathogenicity of *V. virens* and its interaction with rice. Several transformation techniques have been successfully applied to *V. virens*. Electroporation, the process in which a strong electric pulse is applied to an organism in order to transiently increase membrane permeability, has the advantages of being rapid and inexpensive. Through electroporation on conidia, an enhanced green fluorescence protein (eGFP)-expressing *V. virens* strain was obtained, which was able to infect rice flowers and form RFS balls [47]. Polyethylene glycol (PEG)-mediated approach is typically more efficient than electroporation and generally yields a higher percentage of stable transformants. Ashizawa and colleagues [38] engineered a GFP-tagged *V. virens* strain via PEG-mediated transformation on protoplasts, and identified the infection route of *V. virens* in rice spikelets with this strain. *Agrobacterium tumefaciens*-mediated transformation (ATMT) is a fast and easy way to transfer foreign DNA into fungal cells. A few reports have recorded the establishment and optimization of ATMT procedure on *V. virens* conidia, and construction of T-DNA mutant libraries [48, 49], which facilitate characterizing virulence factors in this pathogen. For example, Yu et al. [48] obtained a T-DNA insertion library with 5600 hygromycin-resistant transformants, and identified 37 mutants with impaired pathogenicity. Targeted gene knockout with PEG- or ATMT-mediated transformation has been tried on *V. virens*; however, the homologous gene replacement frequency was very low [50]. Very recently, Liang and colleagues [51] established a CRISPR-Cas9 system, which remarkably increased the frequency of homologous gene replacement. The knockout efficiency could be as high as 50–90% for some *V. virens* genes.

4.3 Genome and pathogenicity of *V. virens*

The availability of *V. virens* genome provides a good basis for charactering its pathogenicity in rice flower. As reported, the genome of UV-8b is approximately 39.4-Mb, encoding 8426 putative proteins [52]. The strain IPU010 possesses a genome of 33.6-Mb, which encodes 6451 predicted proteins [53]. Genome analysis reveals that *V. virens* is evolutionarily closest to the entomopathogen *Metarhizium spp.*, suggesting host jumping from animal kingdom to plant kingdom [52]. Moreover, genome information provides evidence supporting that *V. virens* specifically infects rice flower and has a biotrophic lifestyle, since genes responsible for secreted proteins and secondary metabolism are enriched, while genes associated with polysaccharide degradation and nutrient uptake are diminished [52]. A web-based protein-protein interactive database for *V. virens-Oryza sativa* interaction has been released, greatly facilitating investigation of *V. virens* pathogenicity [54]. Putatively, 628 secreted proteins are encoded by *V. virens* genome, 193 of the secreted proteins are predicted to be effectors [52].

Effectors are powerful weapons possessed by pathogens to manipulate host immune system and metabolisms for successful colonization. Characterizing their roles is important for understanding pathogen-host interactions. In *V. virens* genome, a number of genes encoding effector proteins, such as UV_1261, UV_2508, and UV_2286, have been identified to suppress *Burkholderia glumae*-induced cell death [52], whilst UV_5823 shows ability to suppress plant RNA silencing [55]. On the contrary, some effectors of *V. virens* could induce cell death or defense response in rice protoplast. For example, UV_44 induces cell death, and this ability relies on the serine peptidase active sites. UV_1423 could be *N*-glycosylated, which affects its ability to trigger cell death [56]. However, so far, no effectors have been characterized to function in the flower infection of *V. virens*.

Nevertheless, several virulence factors have been identified in *V. virens*. UvSUN2 is a SUN domain protein; loss of function of this factor results in inability of infecting rice flower, as well as abnormal stress responses and mycelium growth [57]. Mutation in *UvPRO1* increases sensitivity to abiotic stresses and attenuates virulence, in addition to impaired growth rate and sporulation [49]. In contrast, a low-affinity iron transporter encoded by *Uvt3277* negatively regulates virulence in *V. virens* [58].

4.4 Host compatible interaction with *V. virens*

Monitoring host responses to *V. virens* infection could help to uncover the pathogenesis of RFS. In an earlier transcriptome study, a series of differentially expressed genes have been identified in a susceptible rice cultivar 93-11 infected with a field *V. virens* isolate [59]. Among them, genes regulated by Ca^{2+} or abscisic acid are down-regulated, while genes regulated by Myb or WRKY transcription factors are up-regulated. *OsSWEET11* and *OsSWEET14*, which may be involved in disease susceptibility [60], are also up-regulated by *V. virens* infection. Specifically, many pollen development associated genes are down-regulated by *V. virens* infection, but not responsive to other biotic and abiotic stresses, suggesting that these genes may play unique roles in rice-*V. virens* interaction [59]. Additionally, several transcriptome analyses on compatible rice-*V. virens* interactions have been reported. Genes involved in hydrolase, transporter, and flower development tend to be down-regulated in susceptible cultivar Huang-Xiu-Zan upon infection [61]. Expression of many defense-related genes such as *PAL* and *PR* genes could be suppressed in susceptible rice cultivars infected with *V. virens* [42, 62].

As a successful pathogen, *V. virens* should have abilities to set up colonization in rice floral organs and acquire abundant nutrients for propagation, in addition to subvert rice immunity. Transcriptome analysis reveals that genes associated with flower opening, such as *ARF6* and *ARF8* homologs, are down-regulated by *V. virens* infection [42]. This may contribute to inhibition of flower opening during RFS pathogenesis [42]. Furthermore, *V. virens* infection causes failure of ovary fertilization. However, a number of grain-filling-specific genes, such as seed-specific starch synthesis related genes and those encoding seed storage proteins, are activated for high expression in *V. virens*-infected rice spikelets [42]. It is suggested that *V. virens* may be able to mimic fertilization and hijack rice grain-filling system for nutrient supply to pathogen growth and RFS ball formation. This finding is further supported by an independent study [63]. Although the underlying mechanism needs further investigation, the observation of *V. virens* activating rice grain filling could provide a promising explanation why mild *V. virens* infection enhances rice yield traits, including grain weight and filled grain number [9]. Identification and characterization of *V. virens* factors that manipulate rice grain filling should be an interesting research area in the future.

5. Resistance of rice

5.1 Sources and inheritance of RFS resistance

Various attempts have been made to screen rice cultivars resistant to RFS. Screening of 186 rice hybrids to RFS resistance was done by Liang and colleagues [64], which identified few hybrids with low disease incidence. They screened the commercial hybrids that had lower rates of diseased panicles and infected florets at Xindu and Qionglai (Sichuan Province, China) in 2011 together with newly registered varieties. Lore et al. [65] evaluated some hybrids and inbred cultivars growing across India for susceptibility/tolerance to RFS. Artificial inoculation of false smut was done by Kaur et al. [66], which identified nine hybrids resistant to RFS among 125 rice genotypes screened. More detailed evaluation of RFS resistance was performed by Huang and colleagues [67]. A total of 843 rice accessions were screened in disease nurseries in 3 years although some of those accessions were planted in different locations and on different dates. Finally, 36 accessions were found to show no disease incidence. A highly susceptible accession Pujiang 6 was identified in this study. Polymorphism analysis determined several resistant accessions which could be used for crossing with Pujiang 6 to construct gene mapping populations [67].

Resistance of genes against *V. virens* has not been identified yet, but numerous efforts have been undertaken to study the inheritance of the resistance. Earlier, using 266 near-isogenic introgression lines derived from susceptible cultivar Teqing and resistant Lemont and natural infection data in the field, Xu et al. [68] identified two RFS resistance-contributing QTLs, *QFsr10* and *QFsr12*, located on chromosome 10 and 12, respectively. Later, the same group further identified 10 QTLs for RFS resistance [69]. Li et al. [70] developed a population of 157 recombinant inbred lines (RILs) from crossing a susceptible landrace Daguandao (*O. sativa* subsp. japonica) and a resistant cultivar IR28 (*Oryza sativa* subsp. indica). Subsequently, different RILs and parents were evaluated following effective artificial inoculation under field conditions. Genetic analysis showed that the RFS resistance was controlled by two major genes with equal effect of 11.41 and polygenes with minor effects. Further work identified seven QTLs for RFS resistance on chromosomes 1, 2, 4, 8, 10, 11, and 12, and the phenotypic variance ranged from 9.8 to 22.5% [71]. The inheritance of RFS resistance in two-line hybrid rice was investigated using natural infection technique. When a moderate susceptible sterile line TGMS 33S was crossed with susceptible restorer lines, the F1 was susceptible to RFS; when TGMS 33S was crossed with resistant restorer lines, 87% of the F1 showed dominant or incomplete dominant inheritance of RFS resistance, the rest showed recessive inheritance [72]. It should be noted that natural infection of RFS varies among fields and seasons, efficient artificial inoculation method is highly recommended to validate the disease phenotype. Although some progresses on screening for resistant materials and determining the resistance inheritance have been achieved, genes responsible for RFS resistance are still unknown.

5.2 Molecular basis of RFS resistance

Since many rice cultivars with high RFS resistance or high susceptibility have been identified, comparative transcriptome analysis is a promising method to mine resistance- or susceptibility-related genes in rice. For instance, time-course RNA-seq was carried out on susceptible cultivar LYP9 and resistant cultivar IR28 upon *V. virens* infection [62]. Data analysis revealed that many defense-related genes were only up-regulated in the resistant cultivar IR28, but not in LYP9. Particularly, phytoalexin biosynthetic pathway genes such as *OsCPS2*, *OsMAS*, and *OsKSL11* were significantly induced in IR28 at early infection stages, indicating that phytoalexins may contribute to rice resistance against RFS. *PR* family genes, such as β-1,3-glucanase and chitinase genes, were specifically up-regulated in IR28, while generally down-regulated in LYP9. Moreover, a chitinase gene cluster was found close to a RFS resistance QTL on chromosome 11 [71], and nine genes in this cluster were activated by *V. virens* infection. These data suggest that chitinase genes are potential candidates of RFS resistance. In addition, genes encoding receptor-like kinases and WRKY transcription factors may also play roles in RFS resistance. Interestingly, the resistant cultivar IR28 seemed to suppress *V. virens* genes that are associated with pathogenicity and fungal reproduction [62]. Another comparative transcriptome study demonstrated that peroxidase and flavin-containing monooxygenase genes, and genes involved in hormone metabolism were regulated differently in resistant and susceptible cultivars in response to *V. virens* infection [61].

Accumulation of H_2O_2 is a typical plant basal defense fighting against pathogen infections [73]. During a compatible interaction between rice and *V. virens*, obvious H_2O_2 accumulation was detected on the lemma and the palea of infected spikelet. H_2O_2 also enriched in the anthers, stamen filaments, and lodicules of the infected spikelets [74]. However, it needs to be further investigated whether H_2O_2 accumulation pattern is different between resistant and susceptible rice cultivars upon *V. virens* infection. A preliminary study described that higher contents of lignin and polyphenolic compounds were detected in spikelets of resistant rice variety Shuijing 3 than in those of susceptible variety 9522 [75], suggesting the role of these secondary metabolites in RFS resistance. To engineer rice resistant to RFS, an elicitor gene *hrf1* from *Xanthomonas oryzae* pv. *oryzae* was ectopically expressed in R109, which is susceptible to RFS. Artificial inoculation and natural infection both supported that *hrf1* conferred high resistance to the RFS pathogen, presumably through enhancing the expression of defense-related genes, including *OsPR1a*, *OsPR1b*, and *PAL* [76].

Based on the current findings that *V. virens* possesses intercellular infection strategy and biotrophic life style, and that resistant rice cultivars show up-regulation of pathogenesis-related genes and accumulation of secondary metabolites upon infection, basal defense of rice should play a dominant role in resistance against *V. virens*.

6. Disease cycle

V. virens attacks rice flowers and forms RFS balls covered with chlamydospores and/or generate sclerotia, which are considered as primary inocula of RFS disease (**Figure 3**). As to the sexual cycle, a large number of sclerotia can be produced when RFS balls develop in autumn [35]. Sclerotia cannot germinate immediately, requiring a dormancy period of 2–5 months at room temperature or 4°C. They overwinter in the field and could survive up to 10 months with maintaining germination ability to generate ascospores under 25°C and high humidity [77, 78]. Even more, sclerotia can survive with high germination rate up to 5 years when stored in a dry environment at 2–4°C. In the next spring, sclerotia start to germinate, and the germination time varies among different sclerotia. Theoretically, a sclerotium could produce up to 21 million ascospores [35]. Although sclerotia are easy to rot in paddy fields under natural conditions, a limited number of sclerotia can still produce plenty of ascospores. This is supported by the fact that ascospores could be trapped 60 cm above ground in paddy fields between May and September, coinciding with rice booting stage and *V. virens* infection time [35]. Ascospores are able to infect rice flowers to form RFS balls [78, 79]. Therefore, it is believed that sclerotia act as primary inocula of RFS and play an important role in the disease cycle.

With regard to the asexual cycle, chlamydospores from RFS balls are easily transmitted by wind and rainfall, and attack developing rice spikelets of late ripening rice cultivars. This is supported by the fact that fresh chlamydospores have high germination rate and could successfully infect rice flowers to form RFS balls [79, 80]. Chlamydospores can overwinter in soil and on dead plants, or on harvested RFS balls and rice seeds, and survive up to several months; however, the germination rate

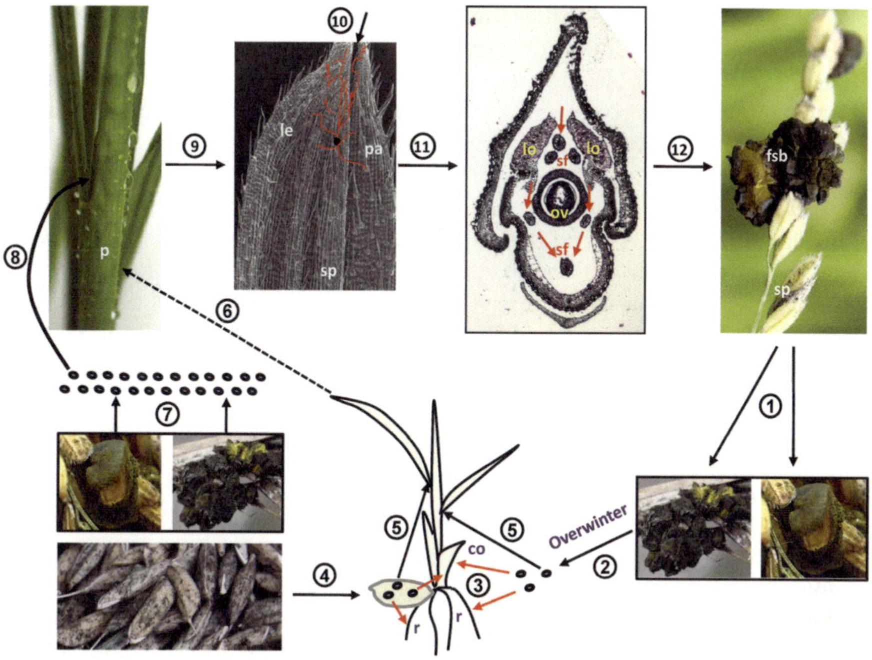

Figure 3.
Disease cycle of rice false smut. Rice false smut balls with chlamydospores and sclerotia are formed in rice spikelets (①), and overwinter in field (②). Next spring, spores in soil (③) and on contaminated rice grains (④) germinate and attack rice roots and coleoptiles when rice seeds are germinating. Hyphae grow intercellularly in roots and coleoptiles, but could not infect seedlings systemically. Instead, hyphae may grow epiphytically on leaf surface or leaf sheath, and reach the external surface between tiller buds at the late vegetative stage (⑤) or even the surface of elongated stems at the heading stage. It is possible that the pathogen hyphae reach the inner space of rice panicles and initiate infection at the late booting stage (⑥). Meanwhile, conidia produced by chlamydospores and/or ascospores from sclerotia (⑦) also initiate attack on rice spikelets in developing panicles (⑧). Spores could firstly germinate on the surface of a spikelet (⑨), and the hyphae extend into the inner space of the spikelet via the gap between the lemma and the palea (⑩). Stamen filaments are the major infection sites for the pathogen (⑪). After successful colonization in floral organs, a large amount of fungal mass are formed and eventually grow into a false smut ball (⑫). The route of infection is indicated by arrows and numbers. Arrows with dotted lines are the steps needing further exploration. Red arrows indicate the main infection sites of the pathogen. Red curve lines represent pathogen hyphae. co, coleoptile; p, panicle; sp, spikelet; le, lemma; pa, palea; sf, stamen filament; lo, lodicule; ov, ovary; and fsb, false smut ball.

decreases rapidly [81]. In the next rice planting season, chlamydospores overwintered in fields and on rice seeds may germinate with hyphae to infect coleoptiles of germinating rice seeds and roots of seedlings [43–45]. Since chlamydospores could not be trapped in fields until RFS balls appear [35], it is unclear how chlamydospore germination time couples with rice booting stage for infecting rice flowers. Studies suggest that coleoptile and root infections may lead to asymptomatic colonization of the pathogen in rice plants at subsequent stages. Sensitive PCR methods have been applied to successfully detect *V. virens* in various tissues of rice before panicle heading [82, 83], suggesting the presence of pathogen in rice plants. Furthermore, colorimetric *in situ* hybridization reveals that *V. virens* mycelia are present on the surface of tiller buds enclosed by young leaf sheaths at vegetative stage, and also on the surface of elongated stems around leaf axils at the heading stage. As *V. virens* infection is not systemic, epiphytic growth could explain how the presence of mycelia in rice plants lasts from the germinating stage to the heading stage of rice [84]. Preset of pathogen mycelia in rice plants especially in leaf sheaths should greatly increase chances of attacking flowers.

Epiphytic growth of *V. virens* is not only found in rice plants, but also detected on leaf surface of various paddy field weeds and on abiotic surfaces (e.g., cellophane and parafilm) [33]. Under wet conditions, *V. virens* conidia are capable of blastogenesis and could produce a large number of secondary conidia on these surfaces in several days. Chlamydospores and ascospores both germinate to produce conidia [78, 81], and the blastogenesis and epiphytic growth greatly increase the amount of inocula under continuing rainy conditions. Therefore, epidemics of RFS disease usually occur when rice booting stage meets with rainy days.

Alternative hosts of a pathogen commonly play an important role in disease cycle. Earlier, paddy field weeds such as *Digitaria marginata* [85], *Panicum trypheron* [86], *Echinochloa crus-galli*, and *Imperata cylindrica* [87] have been reported as alternative hosts of *V. virens*. However, a recent survey demonstrated that infection in these potential alternative hosts is very rare in nature [88]. Still, the presence of *V. virens* in weeds as confirmed by PCR detection [82] and epiphytic colonization on weed leaf surface [33] suggests that paddy field weeds contribute to RFS disease cycle in an unconventional way.

7. Disease control

In recent years, the RFS disease has become a severe threat to rice production due to its epidemics. In order to minimize direct economic loss, suitable management practices have to be made to manage the disease. Breeding and utilization of resistant cultivar is the most effective and economical way to control RFS disease and ensure the high yield of rice. Attempts have been made to identify sources of resistance against *V. virens* (see above). As inheritance of RFS resistance is not well understood, breeding for resistant rice is hindered. Late ripening rice cultivars with large panicle and high grain density are prone to RFS and should be carefully chosen for wide application.

Culture managements have been studied to reduce incidence of RFS. Early planted rice has less RFS balls rather than the late planted rice. Excess application of nitrogenous fertilizer should be avoided. Since high rate of nitrogen increases the disease incidence, sensible use of nitrogen is recommended. Fertilizer ratio is often a reasonable parameter for growers to adjust, so as to enhance the stress tolerance of rice plants, and ultimately reduce the RFS incidence. Field ridges and irrigation channels should be kept clean to eliminate alternative hosts. Conservation tillage and furrow irrigation have some effects on suppressing the disease index [2, 89]. Using suitable plant spacing and utilizing uncontaminated rice seeds are also recommended.

Chemical control, i.e., fungicide application, can be effective but is often not economical and environment-friendly. Using fungicides with high efficiency, low toxicity, and low residue is currently the best choice to control RFS disease. Fungicides, such as Wenquning (a suspension of *Bacillus subtilis* in a solution of validamycin), cuproxat SC, simeconazole, tebuconazole, difenoconazole, and hexaconazole, are effective to reduce RFS disease incidence [64, 90, 91]. It is noteworthy that the timing of spraying fungicides is critical. Application of fungicides after panicle heading should be prevented, as the pathogen infects rice flowers at late booting stage and already successfully colonizes the inner floral organs after heading. As supporting evidence, simeconazole is found to be more effective against RFS when applied 3 weeks before rice heading [92].

8. Disease assay

8.1 Natural infection

To evaluate RFS sensitivity of rice under natural infection, several classification standards of disease incidence have been reported. For example, in 1996, the

International Rice Research Institute (IRRI) [93] classified RFS into 6 scales based on incidence of severely infected tillers or infected spikelets, i.e. 0, no incidence; 1, less than 1%; 3, 1–5%; 5, 6–25%; 7, 26–50%; and 9, 51–100%. Later on, Tang and colleagues [94] established a new classification standard, and developed Disease Index to determine RFS incidence. The classification standard was based on aspect ratio and 100-weight of RFS ball, grain weight, seed setting rate, and yield loss of single diseased panicle. Six scales were classified: 0, no RFS ball; 1, one RFS ball; 2, two RFS balls; 3, 3–5 RFS balls; 4, 6–9 RFS balls; and 5, ≥10 RFS balls. Disease index = $\sum$ (Disease scale value × Diseased plant number)/(Total plant number × Highest disease scale) × 100. Note that only the highest disease scale value is adopted for each plant. This classification standard has been widely applied in recent studies [70–72]. When using natural infection method, disease incidence should be evaluated for multiple years at multiple locations, with multiple sowing dates.

8.2 Artificial inoculation

Due to uncertainty of environmental conditions under natural infection, a high efficient artificial inoculation method is desired for evaluating *V. virens* pathogenicity and rice resistance. As the pathogen specifically infects rice stamen filaments at specific rice stage to cause disease, it is difficult and complicated to optimize an efficient inoculation system. Parameters, such as inocula type, inoculation time and method, incubation conditions after inoculation, and so on, should be considered [42, 95–98]. To date, a number of studies conclude an efficient inoculation method under controlled conditions: first, culture *V. virens* in PSA at 28°C until white colony grows large enough for inoculating into PSB. Usually, 4–8 plugs of mycelia with around 6 mm diameter each are needed, and incubated in PSB at 28°C in dark, 110–150 rpm for 5–7 days. Second, a mixture of mycelia and conidia is blended as inocula, of which the conidia concentration is adjusted to around 10^6 conidia/mL with 4% potato juice. Third, at late booting stage of rice (5–7 days before heading), inocula are injected into panicles with a syringe until the inocula drip out (**Figure 4**). Fourth, the inoculated rice plants are kept at 25°C and 95% relative humidity for 5 days, and then moved to 28°C with relative humidity over 75%. Around 4 weeks post inoculation, disease incidence could be recorded. High RFS incidence (90–100%) has been obtained on susceptible rice cultivars such as Pujiang 6, Yueyou 938, and so on [42, 97]. It should be noted that the artificial inoculation method needs to be modified when applied to different *V. virens* isolates and rice cultivars. For example, the highest RFS incidence is achieved when inoculation is carried out 3–5 days before heading for Yueyou 938 [97], and that is 5–7 days before heading for Pujiang 6 [42]. The disease symptom progression also varies among different *V. virens*-rice combinations under different post-inoculation conditions. As for *V. virens* PJ52-rice Pujiang 6 interaction under artificial inoculation conditions, no obvious symptom could be found through 1 dpi to 5 dpi. At 9 dpi, white fungal biomass can be seen with the naked eye. The fungal biomass enlarges and protrudes out of rice spikelets as early as 13 dpi, and large RFS balls appear at 17 dpi (**Figure 4**).

The above-mentioned classification standard and Disease Index [94] can also be applied to evaluation of disease incidence under artificial inoculation. Alternatively, the following method can be adopted when the disease scale is reaching the highest (i.e., scale 5, ≥10 RFS balls) for each plant. This situation is often encountered when using susceptible rice cultivars to evaluate *V. virens* pathogenicity. Due to high variation of the disease incidence for RFS pathosystem, at least 100 panicles from at least 30 rice plants are recommended to be inoculated. At around 4 weeks post inoculation, each inoculated panicle is collected for counting the number of RFS balls and the number of total spikelets. The number of RFS *b*alls *p*er inoculated *p*anicle

Figure 4.
Artificial inoculation of rice false smut pathogen. (A) Inocula of V. virens are artificially injected with a syringe into a rice panicle at the late booting stage (5–7 days before heading). (B)–(H) Symptom development on rice spikelets after artificial inoculation. No obvious symptom is seen at 1 dpi (day post inoculation) and 5 dpi. White fungal mass could be detected in inner space of the infected spikelets at 9 dpi (E). The fungal mass grows larger at 13 dpi (G), and at 17 dpi eventually forms a ball-shape colony, called false smut ball (H). Scale bar, 0.5 cm. Image (A) is courtesy of Dr. Junjie Yu from Jiangsu Academy of Agricultural Sciences, China.

is recorded as BPP. Percentage of diseased spikelets (PDS) for each inoculated panicle is calculated as: PDS = 100 × number of diseased spikelets/total number of spikelets. To compare pathogenicity among different *V. virens* isolates or resistance/susceptibility among different rice cultivars, BBP and PDS values are recommended to be presented as box-plot. Statistical analysis such as the *ANOVA* test is required to calculate the significance of difference among BBP or PDS datasets.

9. Future aspects

RFS is an emerging disease threatening the production safety of rice grains worldwide. Great progresses have been made to understand the RFS pathogen and its interaction with rice. However, many important questions are yet to be addressed. How much the mycotoxins produced by *V. virens* are contaminated in cooked rice and livestock feed? Whether mycotoxins play a role in pathogenicity? Are there any RFS resistance genes and how do they mediate defense against *V. virens*? How does *V. virens* activate rice grain filling system to hijack nutrient supply for the formation of RFS ball? In addition, an efficient inoculation method mimicking natural infection is highly desired for basic research and accurate evaluation of rice resistance.

Acknowledgements

We thank Drs Dongwei Hu (Zhejiang University) and Junjie Yu (Jiangsu Academy of Agricultural Sciences) for kindly providing images as indicated in figure legends. We apologize to the colleagues whose work could not be included in this book chapter due to space limitation. This work was supported by grants from the National Natural Science Foundation of China (grant no. 31501598 and 31772241) and Key Projects of Sichuan Provincial Education Department.

Conflict of interest

The authors declare no conflict of interest.

Author details

Wen-Ming Wang*, Jing Fan and John Martin Jerome Jeyakumar
Rice Research Institute and Key Lab for Major Crop Diseases, Sichuan Agricultural University, Chengdu, China

*Address all correspondence to: j316wenmingwang@sicau.edu.cn

References

[1] Kabir MS, Salam MU, Chowdhury A, Rahman NFR, Iftekharuddaula KM, Rahman MS, et al. Rice vision for Bangladesh: 2050 and beyond. Bangladesh Renal Journal. 2015;**19**:1-18. DOI: 10.3329/brj.v19i2.28160

[2] Brooks SA, Anders MM, Yeater KM. Effect of cultural management practices on the severity of false smut and kernel smut of rice. Plant Disease. 2009;**93**:1202-1208. DOI: 10.1094/PDIS-93-11-1202

[3] Fan J, Yang J, Wang YQ, Li GB, Li Y, Huang F, et al. Current understanding on *Villosiclava virens*, a unique flower-infecting fungus causing rice false smut disease. Molecular Plant Pathology. 2016;**17**:1321-1330. DOI: 10.1111/mpp.12362

[4] Ladhalakshmi D, Laha GS, Singh R, Karthikeyan A, Mangrauthia SK, Sundaram RM, et al. Isolation and characterization of *Ustilaginoidea virens* and survey of false smut disease of rice in India. Phytoparasitica. 2012;**40**:171-176. DOI: 10.1007/s12600-011-0214-0

[5] Jecmen AC, TeBeest DO. First report of the occurrence of a white smut infecting rice in Arkansas. Journal of Phytopathology. 2015;**163**:138-143. DOI: 10.1111/jph.12263

[6] Tanaka E, Ashizawa T, Sonoda R, Tanaka C. *Villosiclava virens* gen. nov., comb. nov., teleomorph of *Ustilaginoidea virens*, the causal agent of rice false smut. Mycotaxon. 2008;**106**:491-501. DOI: 10.1136/ard.2004.020636

[7] Chib HS, Tikoo ML, Kalha CS, Gupta BR, Singh SK, Raina PK. Effect of false smut on yield of rice. Indian Journal of Mycology & Plant Pathology. 1992;**22**:278-280. DOI: 10.7668/hbnxb.2014.05.001

[8] Jiang Z, Jiang Z, Jiang Z, Zhang G. Effect of false smut ball number on yield of rice. Northern Rice. 2009;**39**:53-54. DOI: 10.3969/j.issn.1673-6737.2009.03.023

[9] Dhua U, Dhua SR, Sahu RK. Precise disease severity assessment for false smut disease of rice. Journal of Phytopathology. 2015;**163**:931-940. DOI: 10.1111/jph.12395

[10] Koiso Y, Li Y, Iwasaki S, Hanaoka K, Kobayashi T, Sonoda R, et al. Ustiloxins, antimitotic cyclic peptides from false smut balls on rice panicles caused by *Ustilaginoidea virens*. Journal of Antibiotics. 1994;**47**:765-773. DOI: 10.7164/antibiotics.47.765

[11] Nakamura K, Izumiyama N, Ohtsubo K, Koiso Y, Iwasaki S, Sonoda R, et al. "Lupinosis"-like lesions in mice caused by ustiloxin, produced by *Ustilaginoieda virens*: A morphological study. Natural Toxins. 1994;**2**:22-28. DOI: 10.1002/nt.2620020106

[12] Zhou L, Lu S, Shan T, Wang P, Sun W, Chen Z, et al. Chemistry and biology of mycotoxins from rice false smut pathogen. In: Melborn BJ, Greene JC, editors. Mycotoxins: Properties, Applications and Hazards. New York, NY, USA: Nova Science Publishers; 2012. pp. 109-130

[13] Luduena RF, Roach MC, Prasad V, Banerjee M, Koiso Y, Li Y, et al. Interaction of ustiloxin A with bovine brain tubulin. Biochemical Pharmacology. 1994;**47**:1593-1599. DOI: 10.1016/0006-2952(94)90537-1

[14] Nessa B, Salam MU, Haque AHMM, Biswas JK, Kabir MS, Macleod WJ, et al. Spatial pattern of natural spread of rice false smut (*Ustilaginoidea virens*) disease in fields. American Journal of Agricultural & Biological Science. 2015;**10**:63-73. DOI: 10.3844/ajabssp.2015.63.73

[15] Koiso Y, Morisaki N, Yamashita Y, Mitsui Y, Shirai R, Hashimoto Y, et al. Isolation and structure of an antimitotic cyclic peptide, ustiloxin F: Chemical interrelation with a homologous peptide, ustiloxin B. Journal of Antibiotics. 1998;**51**:418-422. DOI: 10.1002/chin.199839219

[16] Wang XH, Wang J, Lai DW, Wang WX, Dai JG, Zhou LG, et al. Ustiloxin G, a new cyclopeptide mycotoxin from rice false smut balls. Toxins. 2017;**9**:54. DOI: 10.3390/toxins9020054

[17] Tsukui T, Nagano N, Umemura M, Kumagai T, Terai G, Machida M, et al. Ustiloxins, fungal cyclic peptides, are ribosomally synthesized in *Ustilaginoidea virens*. Bioinformatics. 2015;**31**:981-985. DOI: 10.1093/bioinformatics/btu753

[18] Wang X, Fu X, Lin F, Sun W, Meng J, Wang A, et al. The contents of ustiloxins A and B along with their distribution in rice false smut balls. Toxins. 2016;**8**:262. DOI: 10.3390/toxins8090262

[19] Fu X, Wang X, Cui Y, Wang A, Lai D, Liu Y, et al. A monoclonal antibody-based enzyme-linked immunosorbent assay for detection of ustiloxin A in rice false smut balls and rice samples. Food Chemistry. 2015;**181**:140-145. DOI: 10.1016/j.foodchem.2015.02.068

[20] Fu XX, Xie RS, Wang J, Chen XJ, Wang XH, Sun WB, et al. Development of colloidal gold-based lateral flow immunoassay for rapid qualitative and semi-quantitative analysis of ustiloxins A and B in rice samples. Toxins. 2017;**9**:79. DOI: 10.3390/toxins9030079

[21] Koyama K, Natori S. Further characterization of seven bis(naphtho-γ-pyrone) congeners of ustilaginoidins, pigments of *Claviceps virens* (*Ustilaginoidea virens*). Chemical & Pharmaceutical Bulletin. 1988;**36**:146-152

[22] Shibata S, Ogihara Y. Metabolic products of fungi. XXIII. Ustilaginoidins. 3. The structure of ustilaginoidins B and C. Chemical & Pharmaceutical Bulletin. 1963;**11**:1576-1578. DOI: 10.1248/cpb.11.1576

[23] Shibata S, Ogihara Y, Ohta A. Metabolic products of fungi. XXII. On ustilaginoidins. (2). The structure of ustilaginoidin A. Chemical & Pharmaceutical Bulletin. 1963;**11**:1179-1182

[24] Lu S, Sun W, Meng J, Wang A, Wang X, Tian J, et al. Bioactive bis-naphtho-gamma-pyrones from rice false smut pathogen *Ustilaginoidea virens*. Journal of Agricultural and Food Chemistry. 2015;**63**:3501-3508. DOI: 10.1021/acs.jafc.5b00694

[25] Sun WB, Wang A, Xu D, Wang WX, Meng JJ, Dai JG, et al. New ustilaginoidins from rice false smut balls caused by *Villosiclava virens* and their phytotoxic and cytotoxic activities. Journal of Agricultural and Food Chemistry. 2017;**65**:5151-5160. DOI: 10.1021/acs.jafc.7b01791

[26] Meng J, Sun W, Mao Z, Xu D, Wang X, Lu S, et al. Main ustilaginoidins and their distribution in rice false smut balls. Toxins. 2015;**7**:4023-4034. DOI: 10.3390/toxins7104023

[27] Koyama K, Ominato K, Natori S, Tashiro T, Tsuruo T. Cytotoxicity and antitumor activities of fungal bis-(naphtho-γ-pyrone) derivatives. Journal of Pharmacobio-Dynamics. 1988;**11**:630-635. DOI: 10.1248/bpb1978.11.630

[28] Tsuchiya T, Sekita S, Koyama K, Natori S, Takahashi A. Effect of chaetochromin A, chaetochromin D, and ustilaginoidin A, bis(naphtho-γ-pyrone) derivatives, on the mouse embryo limb bud and midbrain cells in culture. Congenital Anomalies.

1987;**247**:245-250. DOI: 10.1111/j.1741-4520.1987.tb00707.x

[29] Cao B, Park H, Joullie MM. Total synthesis of ustiloxin D. Journal of the American Chemical Society. 2002;**124**:520-521. DOI: 10.1021/ja017277z

[30] Li P, Evans CD, Wu Y, Cao B, Hamel E, Joullié MM. Evolution of the total syntheses of ustiloxin natural products and their analogues. Journal of the American Chemical Society. 2008;**130**:2351-2364. DOI: 10.1021/ja710363p

[31] Fu R, Yin C, Liu Y, Ding L, Zhu J, Zheng A, et al. The influence of nutrient and environmental factors on mycelium growth and conidium of false smut *Villosiclava virens*. African Journal of Microbiology Research. 2013;7:825-833. DOI: 10.5897/AJMR2012.2293

[32] Kim KW, Park EW. Ultrastructure of spined conidia and hyphae of the rice false smut fungus *Ustilaginoidea virens*. Micron. 2007;**38**:626-631. DOI: 10.1016/j.micron.2006.09.006

[33] Fan J, Guo X-Y, Huang F, Li Y, Liu Y-F, Li L, et al. Epiphytic colonization of *Ustilaginoidea virens* on biotic and abiotic surfaces implies the widespread presence of primary inoculum for rice false smut disease. Plant Pathology. 2014;**63**:937-945. DOI: 10.1111/ppa.12167

[34] Fu RT, Ding L, Zhu J, Li P, Zheng AP. Morphological structure of propagules and electrophoretic karyotype analysis of false smut *Villosiclava virens* in rice. Journal of Microbiology. 2012;**50**:263-269. DOI: 10.1007/s12275-012-1456-3

[35] Yong ML, Deng QD, Fan LL, Miao JK, Lai CH, Chen HM, et al. The role of *Ustilaginoidea virens* sclerotia in increasing incidence of rice false smut disease in the subtropical zone in China. European Journal of Plant Pathology. 2018;**150**:669-677. DOI: 10.1007/s10658-017-1312-8

[36] Wang YQ , Li GB, Gong ZY, Li Y, Huang F, Fan J, et al. Stachyose is a preferential carbon source utilized by the rice false smut pathogen, *Villosiclava virens*. Physiological & Molecular Plant Pathology. 2016;**96**:69-76. DOI: 10.1016/j.pmpp.2016.09.003

[37] Wang W, Chen Z. Biological characterization of *Ustilaginoidea virens* with different pathogenecity. Southwest China Journal of Agricultural Sciences. 2013;**26**:2301-2306

[38] Ashizawa T, Takahashi M, Arai M, Arie T. Rice false smut pathogen, *Ustilaginoidea virens*, invades through small gap at the apex of a rice spikelet before heading. Journal of General Plant Pathology. 2012;**78**:255-259. DOI: 10.1007/s10327-012-0389-3

[39] Tang YX, Jin J, Hu DW, Yong ML, Xu Y, He LP. Elucidation of the infection process of *Ustilaginoidea virens* (teleomorph: *Villosiclava virens*) in rice spikelets. Plant Pathology. 2013;**62**:1-8. DOI: 10.1111/j.1365-3059.2012.02629.x

[40] Yong ML, Fan LL, Li DY, Liu YJ, Cheng FM, Xu Y, et al. *Villosiclava virens* infects specifically rice and barley stamen filaments due to the unique host cell walls. Microscopy Research and Technique. 2016;**79**:838-844. DOI: 10.1002/jemt.22710

[41] Rong N, Yong M, Ying XU, Hu DW. Relationship between ultrastructure of cell walls of rice spikelets and infection specificity of *Villosiclava virens*. Acta Botanica Boreali-Occidentalia Sinica. 2017;**37**:1-7. DOI: 10.7606/j.issn.1000-4025.2017.01.0001

[42] Fan J, Guo XY, Li L, Huang F, Sun WX, Li Y, et al. Infection of *Ustilaginoidea virens* intercepts rice seed

formation but activates grain-filling-related genes. Journal of Integrative Plant Biology. 2015;**57**:577-590. DOI: 10.1111/jipb.12299

[43] Ikegami H. Studies on the false smut of rice X. Invasion of chlamydospores and hyphae of the false smut fungus into rice plants. Research Bulletin of the Faculty of Agriculture, Gifu University. 1963;**18**:54-60

[44] Schroud P, TeBeest DO. Germination and infection of rice roots by spores of *Ustilaginoidea virens*. AAES Research Series. 2005;**540**:143-151

[45] Prakobsub K, Ashizawa T. Intercellular invasion of rice roots at the seedling stage by the rice false smut pathogen, *Villosiclava virens*. Journal of General Plant Pathology. 2017;**83**:358-361. DOI: 10.1007/s10327-017-0734-7

[46] Yong M, Liu Y, Chen T, Fan L, Wang Z, Hu D. Cytological studies on the infection of rice root by *Ustilaginoidea virens*. Microscopy Research & Technique. 2018;**81**:389-396. DOI: 10.1002/jemt.22990

[47] Tanaka E, Kumagawa T, Tanaka C, Koga H. Simple transformation of the rice false smut fungus *Villosiclava virens* by electroporation of intact conidia. Mycoscience. 2011;**52**:344-348. DOI: 10.1007/s10267-011-0115-6

[48] Yu M, Yu J, Hu J, Huang L, Wang Y, Yin X, et al. Identification of pathogenicity-related genes in the rice pathogen *Ustilaginoidea virens* through random insertional mutagenesis. Fungal Genetics and Biology. 2015;**76**:10-19. DOI: 10.1016/j.fgb.2015.01.004

[49] Lv B, Zheng L, Liu H, Tang JT, Hsiang T, Huang JB. Use of random T-DNA mutagenesis in identification of gene *UvPRO1*, a regulator of conidiation, stress response, and virulence in *Ustilaginoidea virens*. Frontiers in Microbiology. 2016;**7**:2086. DOI: 10.3389/fmicb.2016.02086

[50] Zheng D, Yi W, Yu H, Xu JR, Wang C. *UvHOG1* is important for hyphal growth and stress responses in the rice false smut fungus *Ustilaginoidea virens*. Scientific Reports. 2016;**6**:24824. DOI: 10.1038/srep24824

[51] Liang Y, Han Y, Wang C, Jiang C, Xu JR. Targeted deletion of the *USTA* and *UvSLT2* genes efficiently in *Ustilaginoidea virens* with the CRISPR-Cas9 system. Frontiers in Plant Science. 2018;**9**:699. DOI: 10.3389/fpls.2018.00699

[52] Zhang Y, Zhang K, Fang A, Han Y, Yang J, Xue M, et al. Specific adaptation of *Ustilaginoidea virens* in occupying host florets revealed by comparative and functional genomics. Nature Communications. 2014;**5**:3849. DOI: 10.1038/ncomms4849

[53] Kumagai T, Ishii T, Terai G, Umemura M, Machida M, Asai K. Genome sequence of *Ustilaginoidea virens* IPU010, a rice pathogenic fungus causing false smut. Genome Announcements. 2016;**4**:e00306-e00316. DOI: 10.1128/genomeA.00306-16

[54] Zhang K, Li Y, Li T, Li ZG, Hsiang T, Zhang Z, et al. Pathogenicity genes in *Ustilaginoidea virens* revealed by a predicted protein-protein interaction network. Journal of Proteome Research. 2017;**16**:1193-1206. DOI: 10.1021/acs.jproteome.6b00720

[55] Fang A. Screening and Identification of the *Usitilaginoidea virens* Effectors that can Induce or Suppress Plant Immunity. Beijing: College of Agriculture and Biotechnology, China Agricultural University; 2014

[56] Fang A, Han Y, Zhang N, Zhang M, Liu L, Li S, et al. Identification

and characterization of plant cell death-inducing secreted proteins from *Ustilaginoidea virens*. Molecular Plant-Microbe Interactions. 2016;**29**:405-416. DOI: 10.1094/MPMI-09-15-0200-R

[57] Yu JJ, Yu MN, Nie YF, Sun WX, Yin XL, Zhao J, et al. Comparative transcriptome analysis of fruiting body and sporulating mycelia of *Villosiclava virens* reveals genes with putative functions in sexual reproduction. Current Genetics. 2016;**62**:575-584. DOI: 10.1007/s00294-015-0563-1

[58] Zheng MT, Ding H, Huang L, Wang YH, Yu MN, Zheng R, et al. Low-affinity iron transport protein Uvt3277 is important for pathogenesis in the rice false smut fungus *Ustilaginoidea virens*. Current Genetics. 2017;**63**:131-144. DOI: 10.1007/s00294-016-0620-4

[59] Chao J, Jin J, Wang D, Han R, Zhu R, Zhu Y, et al. Cytological and transcriptional dynamics analysis of host plant revealed stage-specific biological processes related to compatible rice-*Ustilaginoidea virens* interaction. PLoS One. 2014;**9**:e91391. DOI: 10.1371/journal.pone.0091391

[60] Chen LQ, Hou BH, Lalonde S, Takanaga H, Hartung ML, Qu XQ, et al. Sugar transporters for intercellular exchange and nutrition of pathogens. Nature. 2010;**468**:527-532. DOI: 10.1038/nature09606

[61] Yang C, Li L, Feng A, Zhu X, Li J. Transcriptional profiling of the responses to infection by the false smut fungus *Ustilaginoidea virens* in resistant and susceptible rice varieties. Canadian Journal of Plant Pathology. 2014;**36**:377-388. DOI: 10.1080/07060661.2014.927925

[62] Han Y, Zhang K, Yang J, Zhang N, Fan A, Zhang Y, et al. Differential expression profiling of the early response to *Ustilaginoidea virens* between false smut resistant and susceptible rice varieties. BMC Genomics. 2015;**16**:955. DOI: 10.1186/s12864-015-2193-x

[63] Song JH, Wei W, Lv B, Lin Y, Yin WX, Peng YL, et al. Rice false smut fungus hijacks the rice nutrients supply by blocking and mimicking the fertilization of rice ovary. Environmental Microbiology. 2016;**18**:3840-3849. DOI: 10.1111/1462-2920.13343

[64] Liang Y, Zhang X, Li D, Huang F, Hu P, Peng Y. Integrated approach to control false smut in hybrid rice in Sichuan Province. China. Rice Science. 2014;**21**:354-360. DOI: 10.1016/S1672-6308(14)60269-9

[65] Lore JS, Pannu PPS, Jain J, Hunjan MS, Kaur R, Mangat GS. Susceptibility of rice hybrids and inbred cultivarsto false smut under field conditions. Indian Phytopathology. 2013;**66**:397-399

[66] Kaur Y, Lore JS, Pannu PPS. Evaluation of rice genotypes for resistance against false smut. Plant Disease Research. 2015;**30**:46-49

[67] Huang F, Li Y, Shi J, Fan J, Xu Y-J, Wang W-M. Screening and polymorphism analysis of rice germplasms for resistance to false smut disease in Sichuan Province. Acta Phytotaxonomica Sinica. 2016;**46**:247-257

[68] Xu J, Xue Q, Luo L, Li Z. Preliminary report on quantitative trait loci mapping of false smut resistance using near-isogenic introgression lines in rice. Acta Agriculturae Zhejiangensis. 2002;**14**:14-19

[69] Zhou Y-L, Xie X-W, Zhang F, Wang S, Liu X-Z, Zhu L-H, et al. Detection of quantitative resistance loci associated with resistance to rice false smut (*Ustilaginoidea virens*) using introgression lines. Plant Pathology.

2014;**63**:365-372. DOI: 10.1111/ppa.12091

[70] Li Y-S, Zhu Z, Zhang Y-D, Zhao L, Wang C-L. Genetic analysis of rice false smut resistance using mixed major genes and polygenes inheritance model. Acta Agronomica Sinica. 2008;**34**:1728-1733. DOI: 10.1016/S1875-2780(09)60007-9

[71] Li Y-S, Huang S-D, Yang J, Wang C-L. Analysis of quantitative trait loci for resistance to rice false smut. Acta Agronomica Sinica. 2011;**37**:778-783. DOI: 10.3724/SP.J.1006.2011.00788

[72] Tang S, Gao D, Chen Y, Li Y, Zhang S, Zhao Z. Inheritance pattern of two-line hybrid rice resistance to false smut. Agricultural Science & Technology. 2014;**15**:557-560

[73] Li Y, Lu YG, Shi Y, Wu L, Xu YJ, Huang F, et al. Multiple rice microRNAs are involved in immunity against the blast fungus *Magnaporthe oryzae*. Plant Physiology. 2014;**164**:1077-1092. DOI: 10.1104/pp.113.230052

[74] Hu M-L, Luo L-X, Li J-Q. Effect of *Ustilaginoidea virens* infection on rice spikelets growth and H_2O_2 variation. Acta Phytopathologica Sinica. 2018;**48**:169-175

[75] Dai G, Zhao J, He Y, Jin S, Nicole M. Histochemical observation on the resistant an susceptible varieties to *Ustilaginoidea virens* (Cooke) Tak and the way of infection of conidia. Acta Phytopathologica Sinica. 2005;**35**:37-42

[76] Shao M, Wu Z, Chen B, Luo Sang C, Li L. Resistance of *hrf1* transgenic rice to *Ustilaginoidea virens*. Chinese Journal of Biological Control. 2008;**24**:335-338

[77] Dong K, Fu S. Sprouting test on sclerotia of rice false smut. Journal of Shenyang Agricultural University. 1989;**20**:359-362

[78] Wang G. The sexual stage of *Ustilaginoidea virens* and the infection process of ascospores on rice. Journal Zhejiang Rural Tech Normal School. 1995;**Z1**:1-7

[79] Ikegami H. Studies on the false smut of rice, IV. Infection of the false smut due to inoculation with chlamydospores and ascospores at the booting stage of rice plants. Research Bulletin of the Faculty of Agriculture, Gifu University. 1960;**12**:45-51

[80] Fan R, Wang Y, Liu B, Zhang J, Wang H, Hu D. The process of asexual spore formation and examination of chlamydospore germination of *Ustilaginoidea virens*. Mycosystema. 2010;**29**:188-192. DOI: 10.1002/anie.196700771

[81] Wang G. Studies on the factors influencing the germination of the chlamydospores of *Ustilaginoidea virens*. Acta Phytopathologica Sinica. 1988;**15**:241-245

[82] Zhou YL, Xie XW, Wang S, Pan YJ, Liu XZ, Yang H, et al. Detection of *Ustilaginoidea virens* from rice field and plants by nested PCR. Journal of Agricultural Biotechnology. 2006;**14**:542-545

[83] Ashizawa T, Kataoka Y. Detection of *Ustilaginoidea virens* in rice panicles before and after heading in the field using nested-PCR technique with species-specific primers. Japanese Journal of Phytopathology. 2005;**71**:16-19. DOI: 10.3186/jjphytopath.71.16

[84] Tanaka E, Kumagawa T, Ito N, Nakanishi A, Ohta Y, Suzuki E, et al. Colonization of the vegetative stage of rice plants by the false smut fungus *Villosiclava virens*, as revealed by a combination of species-specific detection methods. Plant Pathology. 2017;**66**:56-66. DOI: 10.1111/ppa.12540

[85] Shetty SA, Shetty HS. An alternative host for *Ustilaginodea virens* (Cooke) Takahashi. IRRI Newsletter. 1985;**10**:11

[86] Shetty SA, Shetty HS. Role of *Panicum-trypheron* in annual recurrence of false smut of rice. Transactions of the British Mycological Society. 1987;**88**:409-411

[87] Atia M. Rice false smut (*Ustilaginoidea virens*) in Egypt. Journal of Plant Diseases and Protection. 2004;**111**:71-82. DOI: 10.1007/BF03356134

[88] Deng Q, Yong M, Li D, Lai C, Chen H, Fan J, et al. Survey and examination of the potential alternative hosts of *Villosiclava virens*, the pathogen of rice false smut, in China. Journal of Integrative Agriculture. 2015;**14**:1332-1337. DOI: 10.1016/S2095-3119(15)61030-9

[89] Brooks SA, Anders MM, Yeater KM. Effect of furrow irrigation on the severity of false smut in susceptible rice varieties. Plant Disease. 2010;**94**:570-574. DOI: 10.1094/pdis-94-5-0570

[90] Ahonsi MO, Adeoti AYA. Evaluation of fungicides for the control of false smut of rice caused by *Ustilaginoidea virens* (Cooke) Tak. Journal of Agricultural Research. 2003;**4**:118-122. DOI: 10.4314/mjar.v4i1.31764

[91] Zhou J, Wang X-S. Field efficacy of 27.12% cuproxat SC against rice false smut *Ustilaginoidea oryzae*. Agrochemicals Research & Application. 2011;**15**:21-22

[92] Tsuda M, Sasahara M, Ohara T, Kato S. Optimal application timing of simeconazole granules for control of rice kernel smut and false smut. Journal of General Plant Pathology. 2006;**72**:301-304. DOI: 10.1007/s10327-006-0288-6

[93] IRRI. Standard Evaluation System for Rice. Manila: Philippines; 1996

[94] Tang C, Gao J, Cao G. Research and application of classification standard of rice false smut. Plant Protection. 2001;**27**:18-21

[95] Zhang JC, Chen ZY, Zhang BX, Liu YF, Lu F. Inoculation techniques used for inducing rice false smut efficiently. Acta Phytopathologica Sinica. 2004;**34**:463-467. DOI: 10.1300/J064v24n01_09

[96] Guo X, Li Y, Fan J, Li L, Huang F, Wang W. Progress in the study of false smut disease in rice. Journal of Agricultural Science and Technology A. 2012;**2**:1211-1217

[97] Jia Q, Lv B, Guo MY, Luo CX, Zheng L, Hsiang T, et al. Effect of rice growth stage, temperature, relative humidity and wetness duration on infection of rice panicles by *Villosiclava virens*. European Journal of Plant Pathology. 2015;**141**:15-25. DOI: 10.1007/s10658-014-0516-4

[98] He H, Sopone W, Yuan J, Yang X, Chen X, Wu S, et al. Biology and artificial inoculation of *Ustilaginoidea virens* (Cooke) Takahashi in rice. African Journal of Microbiology Research. 2015;**9**:821-830. DOI: 10.5897/AJMR2014.6736

Chapter 8

Disease Resistance and Susceptibility Genes to Bacterial Blight of Rice

Tariq Mahmood and Frank F. White

Abstract

Rice (*Oryza sativa* L.) is a valuable resource for understanding the complex processes controlling yield and value-added traits. Bacterial blight (BB) is a vascular disease of rice, caused by strains of *Xanthomonas oryzae* pv. *oryzae* (*Xoo*) and provides insight, both practical and basic, into the concepts of susceptibility and resistance. Basic knowledge has been empirically and, more recently, intentionally exploited for broad and durable resistance to the disease. Bacterial blight involves representatives of most classes of resistance genes (*R* genes) and pathways for basal plant immunity. The study of BB also revealed novelties not observed in other models, possibly due to the long history of rice cultivation and the constant disease pressure. Conspicuous are the recessive *R* genes that target the notorious type III Transcription Activator-like effectors (TALes) of *Xoo*. Results indicate that pathogen and host are currently in a battle over a small patch of ground involving TALes function. At the same time, analyses of rice disease physiology are adding to a growing body of knowledge for plant disease processes and to how these processes are intertwined with disease susceptibility. The basic processes of BB present rich targets for the rapid advances in genome editing.

Keywords: *Xanthomonas oryzae* pv. *oryzae*, rice, recessive resistance, TAL effector, genome editing, CRISPR

1. Introduction

World population is expected to rise beyond 9 billion by 2050 [1]. Rice (*Oryza sativa*) is a staple food crop world-wide, providing about one fifth of the calories consumed by humans [2]. In particular, rice accounts for 35–75% of the calories consumed by more than 3 billion in Asian countries alone and planted on approximately 154 million hectares land annually [3]. Crop protection and food security go hand in hand, and breeding for resistance against crop diseases remains the essential ingredient for food security. Due to the labor-intensive nature of breeding, integrated disease control is often reduced to mere chemical control, leaving the very purpose of this environment-friendly approach in limbo. Advances in molecular tools in crop breeding, however, makes breeding an increasing sustainable effort in staying ahead of pathogen adaptation [4]. Bacterial blight (BB) of rice is a widespread vascular disease caused by *Xanthomonas oryzae* pv. *oryzae*

IntechOpen

(*Xoo*). Epidemics can severely reduce grain yield due to collapse of the entire crop [5]. BB was first characterized in the late nineteenth century [6]. Introduction of resistance (*R*) genes into rice cultivars is considered as the best option for *Xoo management*. A total of 42 *R* genes have been identified in rice against *Xoo, and the number continues to grow* [7–9]. Due to co-evolution and selection pressure between *Xoo* and rice, these *R* genes are selective in their efficiency against specific *Xoo* strains or races, which are sets of strains that share incompatibility on defined sets of *R* genes [10].

2. Post genomic era and rice grain protection

Advancements in genomics, referring here to DNA and RNA analyses, is as beneficial to crop protection as is to other discipline of biology. Rice MetaSysB, an open source which provides detailed information about BB-responsive genes, is based on the global expression analysis. The database provided 7475 unique genes and 5375 simple sequence repeats, which were responsive to *Xoo* in rice [11]. Such information is based on the compatible and incompatible rice-*Xoo* interactions. In another example, 454 and 498 differentially expressed genes were reported as exemplified by the incompatible and compatible rice-*Xoo* interactions, respectively, using cDNA microarray [12]. Genomics also provides functional information of genes up- and downstream of candidate resistance genes in the defense signal pathway, as is done in near-isogenic rice lines introgressed with *Xa39*, an as yet uncharacterized BB resistance gene [13].

Multiple rice and *Xoo* genomes have been sequenced, either in draft or complete form [14–23], paving the way to identify functional connections between host and pathogen genes. The functional validation of the candidate genes is helping develop new rice varieties by introduction of the gene of interest through traditional breeding, marker assisted breeding, or genetic engineering approaches [3]. BB disease resistance is overcome by the emergence of more virulent strains of *Xoo*. Whole genome sequencing of 100 *Xoo* strains from India revealed that these strains were distinct from African and US *Xoo* strains [24]. Based on the reaction towards ten major resistance genes of rice, 46 out of the 100 strains were grouped into 11 pathotypes [24].

3. The genetic context of rice-*Xoo* interaction

Many BB-resistance genes in modern rice germplasm were selected long before the concepts of modern plant breeding were established, and a rich assortment of major dominant and recessive *R* genes has been identified by genetic and molecular studies (**Table 1**).

Perhaps the best known of these genes, *Xa21* represents the receptor kinase (RLK) class of *R* genes. *Xa21* was originally introgressed into rice from the related species *O. longistaminata* and confers resistance to a broad range of *Xoo* strains [25]. *Xa26*, another cloned member of RLK gene family, also confers broad resistance with a somewhat different strain profile [26]. The cognate elicitor for Xa21 has been reported [27]. However, for *Xa26* has not been identified.

RLKs play a central role in disease immunity pathways in plants, largely via the characterization of the bacterial flagellin receptor FLS2 and the related receptor EFR in *Arabidopsis* [28, 29]. A typical RLK consists of an extracellular receptor domain comprising of leucine-rich repeats (LRRs), a transmembrane domain, and an intracellular kinase domain [30]. As a class, RLKs have great potential for

Gene	Class	Comments	Cognate elicitor/ effector	Ref
Xa21	RLK[1]	extracellular, membrane and intracellular domains; kinase; broad resistance	RaxX	[25, 27]
Xa26	RLK	similar to *Xa21*; same locus as *Xa3*; broad resistance	Unknown	[26]
Xa1, *Xo1*	NBS-LRR[2]	cytoplasm; narrow resistance	Multiple TALes	[31–33]
Xa4	WAK[3]	narrow	unknown	[40]
Xa27, *Xa23*, *Xa10*	TAL effector inducible	membrane and cell wall; novel protein; broad resistance	AvrXa27, AvrXa23, AvrXa10	[37–39]
xa5	Missense mutant of *TFIIAγ5*; small subunit of TFIIA transcription factor complex	nuclear; broad resistance	TALe interference	[51, 53, 54]
xa13	promoter mutants of *OsSWEET11*; nodulin 3 family	membrane; unresponsive to PthXo1	PthXo1	[42, 47]
xa25, *OsSWEET13^{Kit}*	promoter mutant of OsSWEET13, nodulin 3 family	TATA box polymorphisms; unresponsive to PthXo2	PthXo2	[44, 52]

[1]*RLK, receptor linked kinase.*
[2]*NBS-LRR, nucleotide binding site, leucine-rich repeat.*
[3]*WAK, wall-associated kinase.*

Table 1.
Cloned R genes to bacterial blight of rice.

enhancing resistance to BB in rice and in other disease complexes of crop plants *Xa21*, *Xa26*, and other RLKs represent genetic components of the pathogen-associated molecular patterns (PAMPs)-triggered immunity (PTI) surveillance pathway in rice. Improvements in the rationale design of RLK receptor specificities, and screening for novel genes in germplasm or wild relatives could lead to general application for broad and durable resistance.

The nucleotide binding site-LRR (NBS-LRR) is another large class of *R* gene, represented in rice toward *Xoo* by *Xa1* and *Xo1* [31–33]. XA1 and XO1 recognize multiple TALe, and *Xoo* strains have adapted TALes, the so-called iTALes, that are truncated and inhibit the function of XA1 and XO1 [32, 34].

Specific TALe-dependent *R* genes governing dominant resistance in rice against *Xoo* are known as executor (*E*) genes. *E* genes are distinct from classical *R* genes, whose transcriptional activation by TALes of *Xoo* trigger immunity, leading to dominant resistance [35]. *Xa27* represents the E genes class of dominant *R* genes and confers broad resistance to BB in rice [36]. Although not expressed in susceptible host, *Xa27* is expressed only upon inoculation with *Xoo* strains harboring the TALe gene *avrXa27* [37]. The protein is localized to apoplastic space, cell membrane and cell wall, and when expressed under a pathogen-nonspecific inducible rice *OsPR1* promoter, conferred constitutive resistance to both compatible and incompatible strains alike [37]. The rice *R* genes *Xa10* and *Xa23* have similar requirements for the transcription activation domain and nuclear localization sequence (NLS) motifs of the corresponding TALes for their induction [38, 39].

Xa4 is the latest and, again, an unusual *R* gene of rice to be characterized. The protein is a wall-associated kinase (WAK) and provides attributes other than enhanced resistance. Rice plants with XA4 are shorter and stiffer in comparison to plants lacking the gene [40]. Xa4 is race-specific, meaning many strains of *Xoo* are compatible on plants with Xa4. How Xa4 functions in resistance is unknown at present.

3.1 *SWEET* genes and recessive resistance

A class of major TALe-dependent susceptibility (*S*) genes for BB in rice encodes sugar transporters, thereby named as SWEET gene family [41]. Specific TALes, referred to as major TALes, transcriptionally activate the corresponding SWEET genes in rice during infection to promote the disease in a gene-for-gene susceptibility manner [42]. Although at least five SWEET genes of the clade III members can function as an *S* gene in BB, only three members are known to be targeted by extant strains of *Xoo* [42–47]. A member of the SWEET gene family, *OsSWEET14*, is induced by multiple distinct TALes, which include AvrXa7, PthXo3, Tal5 and TalC and are present in strains of different geographic origins and genetic lineages [43, 45, 46]. Similarly, PthXo2 drives *OsSWEET13* expression in the susceptible rice variety IR24 [44], and *OsSWEET11* is induced by the cognate PthXo1 [42]. The typical TALe possesses a central repetitive domain, a nuclear localization signal domain, and a transcription activation domain. The repetitive domain is responsible for binding of the TALe to a sequence motif called the effector binding element (EBE), which is commonly located in the promoter region of the respective *S* gene.

Mutated *S* gene alleles are proposed to be potentially more durable than dominant *R* genes [48, 49]. Identifying the promoter variant alleles of major *S* genes has been proposed in breeding for BB resistance [42, 47, 50–53]. Recessive resistance is due to the cognate TALe cannot bind to the promoter variants of the *S* gene. The gene *xa13*, for example, is a recessive resistance insertion allele of 14.8 kb DNA fragment in the promoter of *OsSWEET11* [42, 47]. *OsSWEET11* encodes a protein related to *MtN3* encoding nodulin 3 (N3) protein of *Medicago truncatula*. The gene was originally named *Os8N3* due to its location on rice chromosome 8 and the similarity to *MtN3* [42]. The critical difference between resistant (*xa13/xa13*) and susceptible plants is the elevated expression of *OsSWEET11* during infection in otherwise susceptible plant genotypes [42]. RNAi-mediated silencing of *OsSWEET11* plants was similarly resistant to *Xoo* strains that are solely dependent on PthXo1 for SWEET induction. Silenced plants, but not promoter variants, showed low pollen viability, corroborating the fact that *Xoo* hijacked otherwise developmentally important genes in rice for pathogenicity [42, 47]. Similarly, the TALe PthXo2 cannot bind to the EBE of *xa25*, a recessive allele of *OsSWEET13*, or the EBE region of *OsSWEET13* in japonica rice cultivars, owing to single nucleotide polymorphisms in the respective EBEs [51, 52].

The gene *xa13* is a naturally occurring allele, actually a series of alleles that protects the plant from a genetic disease vulnerability in the plant developmental pathways [42, 47]. However, *xa13* is not a broad resistance provided in comparison to *Xa21*, *Xa27* and *xa5*), and many strains from China, Philippines, Japan and Korea are compatible on *xa13* lines [51]. Compatibility is derived by acquisition of major TALes that target alternative SWEET promoters [43]. As yet, not major TALe has been identified that replaces PthXo1 for *OsSWEET11* expression.

The gene *xa5* also affects TALe-dependent function but does not act at a specific SWEET gene. The recessive allele encodes a variant of the γ or small subunit of the transcription factor TFIIA [54, 55], which confers broad resistance. The gene differs from the susceptible allele by a single codon substitution of valine at position 39 to glutamic acid. TFIIA, consists of α, β, and γ subunits, and is involved in stabilizing the binding of the TATA box binding protein complex (TFIID) to the TATA box of gene promoters. The TFIIA components are highly conserved across the eukaryotes. Rice has two loci for *TFIIAγ*-one gene is on chromosome 5 (*TFIIAγ5, xa5*) and another on chromosome 1 (*TFIIAγ1*) [54]. The proteins are closely related but not identical. *xa5* provides broad BB resistance and functions in inhibiting TALe function [51, 56]. However, *xa5* is not effective against strains with the TALe PthXo1 [51].

Perhaps not all *SWEET S* genes are known or are not always induced in disease by *Xoo*. The Indian strain IX-80 was virulent but did not induce any known *SWEET* gene [57], suggesting an adaptation by the *Xoo* to relieve dependency on *SWEET* gene family. On the other hand, IX-80 remains TALe-dependent as the strain was not compatible on IR53 (*xa13/xa13, xa5/xa5*), a gene combination that blocks the xa5-compatible PthXo1 and all other major TALes at *OsSWEET14* and *OsSWEET14* [51].

4. Implication of interactions between TALes and the corresponding host genes

Due to the large reservoir of TALes in each strain of *Xoo* and the diverse roles of TALes in pathogenesis, the BB of rice represents an excellent plant/pathogen system for studying the biology of TALes. The apparent reason for the broad activity of *Xa27* and *Xa23* is the presence of the cognate TALes *avrXa27* and *avrXa23* in a large number of strains from southeast Asia, including Korea, China, Japan and the Philippines [37, 39]. On the other hand, the loss of *avrXa27*, *avrXa23*, or *avrXa10*, for that matter, does not appear to have an apparent fitness cost to the pathogen, and populations of *Xoo* may lose *avrXa27* if *Xa27* is widely deployed [37–39]. AvrXa7 is an important virulence factor for some strains of *Xoo*, and strains with AvrXa7 are incompatible on rice lines harboring the *Xa7*. In this case, loss of *avrXa7*, which is a major TALe for *OsSWEET14*, may result in strains that are weakly virulent or, essentially, nonpathogenic, if no other SWEET inducing TALes are present [43, 58]. A variety of other TALe genes are present in *Xoo* populations that can restore full virulence to strains missing *avrXa7* [59]. Evasion of *Xa7*-mediated resistance is possible by loss of the gene, rearrangement of the central repeats or recombination among different TALe genes [60, 61]. However, despite rapid adaptation of bacteria by genetic changes and gene flow, field studies in the Philippines indicated that deployment of *Xa7* was durable in test plots for more than 10 years [62]. Therefore, strains may have other limitations due to geographical location or rice genotype. Nevertheless, pyramiding broadly effective *R* genes with cognate TALes that are wide-spread in the pathogen populations should provide a degree of broad and durable resistance.

In the case of *xa13*, induction of the dominant allele *SWEET11* is mediated by the TALe PthXo1 [42]. However, strains of *Xoo* that solely rely on PthXo1 cannot induce *xa13* allele, and rice homozygous for *xa13* is symptomless. *xa13*-dependent recessive resistance is phenotypically and qualitatively different from resistance provided by the dominant *R* gene *Xa7* [42, 63]. Quantitatively, however, resistance mediated by *xa13* and *Xa7* are approximately equal with respect to bacterial growth and lesion length [42, 58, 64]. *Xa7* resistance is the result of the presence of the appropriate AvrXa7 in the pathogen and dominant, while *xa13* resistance is dependent on the absence of an effective virulence factor and recessive. The mechanism of XA7 mediated resistance is as yet unknown.

Type III effectors, in general, are hypothesized to interfere with host defense and defense signaling mechanisms. Strains of *Xoo* have other type III effectors, differing from TALes, and, therefore, not entirely dependent on TALes for suppression of host defenses [65]. *Xoo* strains lacking major TALes are still capable of causing water-soaking, if syringe inoculated, which is in contrast to type III secretion system (Hypersensitive reaction/pathogenicity or Hrp$^-$) mutants. Hrp$^-$ mutant strains are incapable of secreting any type III effectors, including TALes, and are virtually symptomless [66]. The mechanism by which SWEET transporters condition susceptibility is unknown. One hypothesis is that the transporters allow cells to leak sucrose, providing the pathogen with nutrients. SWEET function may interfere with normal plant defense functions or, possibly, allow transport of other nutrients or disease promoting compounds [41]. However, little empirical evidence for the nutrition model exists at present.

Sequencing of *Xoo* genomes has revealed the full complement of TALes is now known [17–23]. The individual TALe genes are distinguishable on the basis of the number of repeats in the central repetitive region and by polymorphisms within each repeat sequence, particularly, at the 12th and 13th codons. Strains of the Asian lineage contain upwards of 16–19 TALe genes in each genome [18]. The large numbers of TALe genes in these species may reflect the evolutionary investment in utilizing the TALes for virulence and are essential, to the ecological niche these bacteria occupy. The maintenance of a large repertoire of TALe genes may increase the frequency of recombination between, and diversity of TALecgenes within the pathogen population [60]. Pathogen may then adapt faster to the changing host genotypes as exemplified by the appearance of *pthXo5,* which avoids Xa7 recognition and appears to be a hybrid between *avrXa7* and *pthXo6* [61].

Not all TALE genes of *Xoo*, however, are just substrates for new major TALEs. Two other TALE genes from PXO99 strain of *Xoo*, in addition to *pthXo1*, contribute to virulence, known to elevate the expression of two host genes distinct from *SWEET11*. PthXo6 elevates the expression of *OsTFX1*, which contributes to approximately 35% of the disease [67]. Many strains induce *OsTFX1*. The gene *pthXo7* of PXO99 elevates the expression of *OsTFIIAγ1* and would appear to be an adaption to host genotypes containing the *xa5* allele of *TFIIAγ5* [67]. However, introduction of *pthXo7* to other strains does not restore full virulence on *xa5*/*xa5* plants and may provide only an incremental fitness benefit [67]. All Asian strains also carry a set of truncated TALes, the inhibitory or iTALes, which function to suppress *Xa1*-mediated resistance [32].

4.1 Executor *R* genes and super promoters

Xa10, *Xa23* and *Xa27* are representatives of the new class of E genes, so-named because the induction of these genes executes a response of programed cell death (PCD) in the host. *Xa10* induced PCD in plant species rice and *N. benthamiana,* and mammalian HeLa cells [38]. No cognate *S* genes for AvrXa10, AvrXa23, or AvrXa27 in compatible host cultivars have been reported, though the presence of AvrXa27 and AvrXa23 in many extant strains of *Xoo* may portend either a defeated function or an unknown cryptic function in *S* gene expression. Nonetheless, *E* genes hold great potential for broad and durable resistance in rice against extant *Xoo* population. A super promoter consisting of multiple EBEs, corresponding to specific TALes in extant population of *Xoo*, have been constructed (**Figure 1**). [68–70]. Addition of multiple EBEs to a pathogen strain specific rice BB resistance gene makes it effective against additional strains of *Xoo*. The EBEs of TALes PthXo1, PthXo6 and Tal9a when conjugated to E gene *Xa27*, showed resistance against PXO99 and a derivative strain lacking *AvrXa27* [68]. A similar scenario was

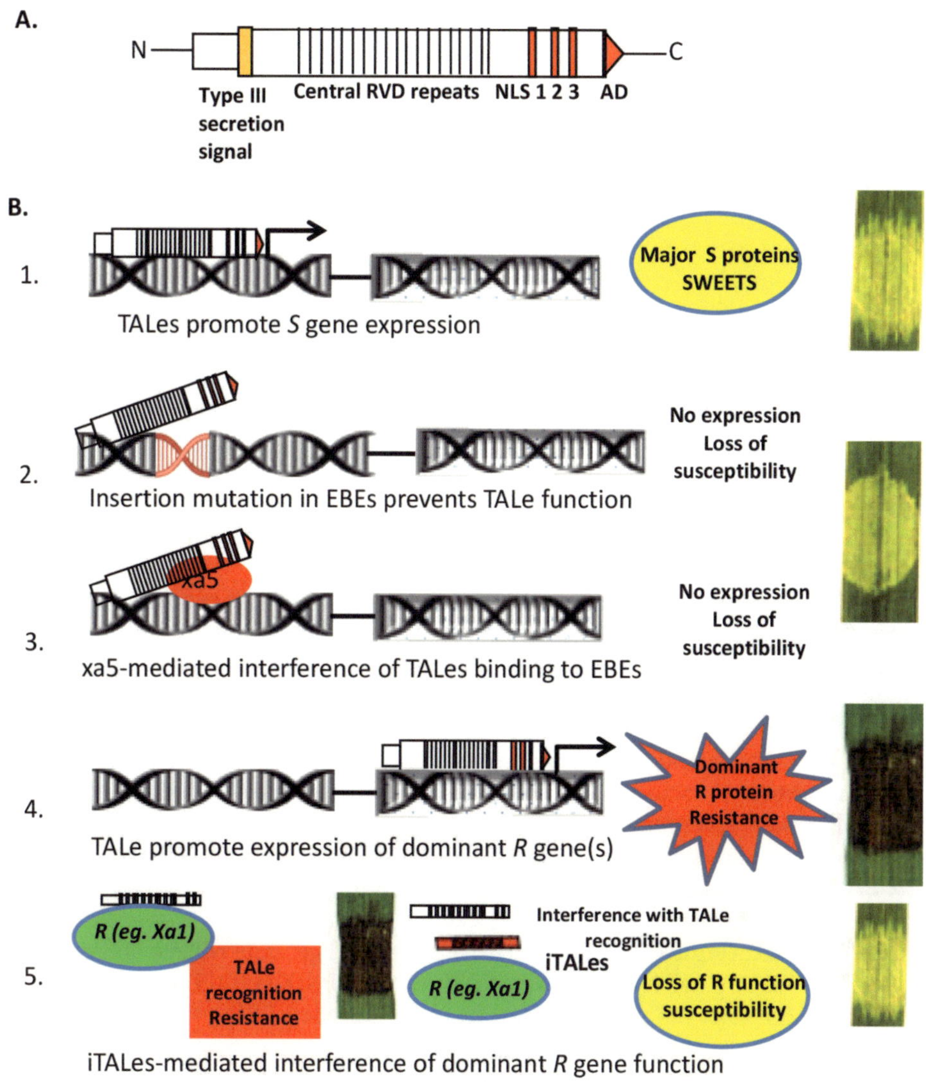

Figure 1.
Xoo TALe-dependent resistance and susceptibility in BB of rice. (A) Schematic of typical TALe from Xoo and (B) five types of TALe interactions affecting outcome of Xoo and rice interaction.

accomplished using E gene *Xa10* [69]. The study suggested that broad-spectrum and potentially durable resistance is possible by stable integration of an E gene engineered in a way to respond to multiple TALes from different strains or even different pathogens. Design of a super promoter, however, needs to be done carefully. Risk that an added EBE might coincidently contain a *cis* regulatory element could induce the E gene expression in response to particular stimuli and cause cell death without challenge by TALes. Amended promoters should be tested thoroughly before deployment.

4.2 Targeted genome regulation and editing

Central to TALe function is the discovery of the DNA recognition cipher of TALEs [71, 72]. The central domain of a TALe, also known as binding domain, consists of variable number of tandem repeats, each consisting 33–35 amino acid residues. The 12th and 13th amino acid residues (known as repeat variable di-residues, RVDs) of each repeat preferentially binds to the respective nucleotides in the EBEs of target gene, such that HD, NG, NI and NN bind to C, T, A, and G, respectively in the effector binding elements (EBEs) of the promoter of a target gene [71–73]. The TALe recognition code allowed custom-engineer of DNA binding domains, also called designer TALes (dTALes), with novel specificity to the user-chosen DNA sequences [74–76]. dTALes provide a useful tool box to transiently activate host genes of interest for their functional analysis and assess the associated effect on host phenotype and physiology during rice-*Xoo* interaction. TALENs are fusions between dTALes and the nuclease domain of restriction enzyme FokI [77–80]. Other C-terminal domains have also been used [81]. Target site recognition and TALEN dimerization triggers a double-strand break (DSB) and generates small random insertions or deletions at the cleavage site, resulting in an edited sequence. CRISPR-Cas editing approaches have circumvented the need to construct dTALes and achieved wide general use, including editing of rice genes [82–84].

5. Prospects for engineered broad and durable resistance in rice to BB

Traditional resistance breeding has identified many useful *R* genes and introgressed the genes into elite cultivars. Further, development of molecular markers allows the pyramiding of multiple genes into single lines. The development of designer TALENs and CRISPR-Cas genome editing brings greater flexibility and rapidity to the development of resistant germplasm. A continuous provision of novel *R* genes in breeding programs is possible. Of course, the adoption and utility of different approaches is dependent on the regulatory climate. Introduction of novel or alien genes may be prohibitive in the foreseeable future. Classification of genome editing techniques will also vary depending on the individual country. In the rice system, our understanding allows numerous approaches for the enhancement of resistance beyond classical breeding. TALe biology, specifically, can be exploited (**Figure 2**). Least intrusive is targeted genome editing of *S* genes. *OsSWEET14* is targeted by unrelated TALEs, AvrXa7, PthXo3, Tal5 and TalC from different *Xoo* strains and which in some cases overlap their EBEs [43, 45, 46]. *OsSWEET14* was made unresponsive to TALEs AvrXa7 and Tal5, when their respective EBEs were mutated using TALENs in otherwise susceptible rice cv. Kitake [85, 86]. Thus, recessive resistance obtained by the genome editing of *OsSWEET14* is expected to be broad and contribute to durability given the apparent few major TALEs in the extant population. Future efforts will be to target all EBE/*S* gene combinations in single elite lines. Fusion of EBEs to a variety of R and E genes has

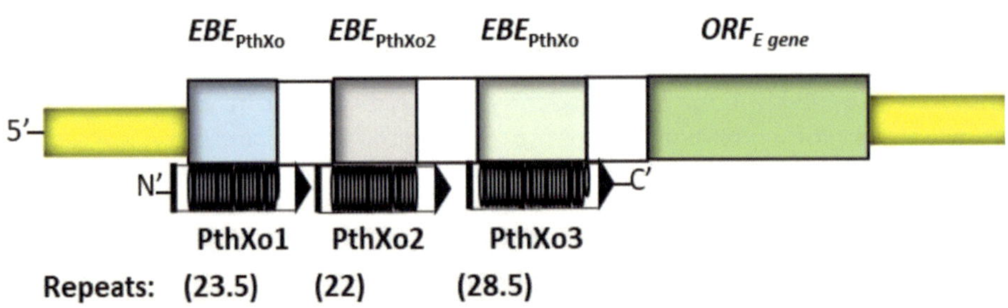

Figure 2.
Super promoters: pyramiding of EBEs of multiple TALEs upstream of an E gene for broad and durable resistance. Subscript of each EBE corresponds to the respective TALEs. Blocks under each EBE represent the respective TALEs with blunt ends as their N termini, and arrowheads as their C-termini flanking the binding repeats in center.

been demonstrated to provide resistance [68, 87]. The functional specificity of an E gene can be broadened by linkage to general inducible defense genes [69, 88]. Approaches are not limited to TALe-associated responses. The RLK immunity receptor EFR from *Arabidopsis* [89, 90], as well as XA21/EFR fusion proteins function in rice [91]. Thus, the sky is the limit for the engineering of broad and durable resistance in rice to BB.

Author details

Tariq Mahmood[1] and Frank F. White[2*]

1 Department of Agriculture, Hazara University Mansehra, Pakistan

2 Department of Plant Pathology, University of Florida, Gainesville, FL, United States

*Address all correspondence to: ffwhite@ufl.edu

References

[1] Béné C et al. Feeding 9 billion by 2050—Putting fish back on the menu. Food Security. 2015;**7**(2):261-274

[2] Kennedy G, Burlingame B. Analysis of food composition data on rice from a plant genetic resources perspective. Food Chemistry. 2003;**80**(4):589-596

[3] Khush GS. What it will take to feed 5.0 billion rice consumers in 2030. Plant Molecular Biology. 2005;**59**(1):1-6

[4] Collard BC, Mackill DJ. Marker-assisted selection: An approach for precision plant breeding in the twenty-first century. Philosophical Transactions of the Royal Society, B: Biological Sciences. 2007;**363**(1491):557-572

[5] Mew T. Current status and future prospects of research on bacterial blight of rice. Annual Review of Phytopathology. 1987;**25**(1):359-382

[6] Reddy A et al. Relationship of bacterial leaf blight severity to grain yield of rice. Phytopathology. 1979;**69**(9):967-969

[7] Ou SH. Rice Diseases. IRRI; 1985

[8] Kim SM et al. Identification and fine-mapping of a new resistance gene, Xa40, conferring resistance to bacterial blight races in rice (*Oryza sativa* L.). Theoretical and Applied Genetics. 2015;**128**(10):1933-1943

[9] Busungu C et al. Identification and linkage analysis of a new rice bacterial blight resistance gene from XM14, a mutant line from IR24. Breeding Science. 2016;**66**(4):636-645

[10] Mishra D et al. Pathotype and genetic diversity amongst Indian isolates of *Xanthomonas oryzae* pv. *oryzae*. PLoS ONE. 2013;**8**(11):e81996

[11] Sureshkumar V et al. RiceMetaSysB: A database of blast and bacterial blight responsive genes in rice and its utilization in identifying key blast-resistant WRKY genes. In: Database. 2019

[12] Li Q et al. Expression profiling of rice genes in early defense responses to blast and bacterial blight pathogens using cDNA microarray. Physiological and Molecular Plant Pathology. 2006;**68**(1-3):51-60

[13] Zhang F et al. Comparative transcriptome profiling of a rice line carrying Xa39 and its parents triggered by *Xanthomonas oryzae* pv. *oryzae* provides novel insights into the broad-spectrum hypersensitive response. BMC Genomics. 2015;**16**:111

[14] RGSP International. The map-based sequence of the rice genome. Nature. 2005;**436**(7052):793

[15] Sun C et al. RPAN: Rice pan-genome browser for ~3000 rice genomes. Nucleic Acids Research. 2017;**45**(2):597-605

[16] Ochiai H et al. Genome sequence of *Xanthomonas oryzae* pv. *oryzae* suggests contribution of large numbers of effector genes and insertion sequences to its race diversity. Japan Agricultural Research Quarterly: JARQ. 2005;**39**(4):275-287

[17] Lee B-M et al. The genome sequence of *Xanthomonas oryzae* pathovar *oryzae* KACC10331, the bacterial blight pathogen of rice. Nucleic Acids Research. 2005;**33**(2):577-586

[18] Salzberg SL et al. Genome sequence and rapid evolution of the rice pathogen *Xanthomonas oryzae* pv. *oryzae* PXO99A. BMC Genomics. 2008;**9**:204

[19] Booher NJ et al. Single molecule real-time sequencing of *Xanthomonas oryzae* genomes reveals

a dynamic structure and complex TAL (transcription activator-like) effector gene relationships. Microbial genomics. 2015;**1**(4):1-22. DOI: 10.1099/mgen.0.000032

[20] Huguet-Tapia JC et al. Complete genome sequence of the African Strain AXO1947 of *Xanthomonas oryzae* pv. *oryzae*. Genome Announcements. 2016;**4**(1). DOI: 10.1128/genomeA.01730-15

[21] Quibod IL et al. Effector diversification contributes to *Xanthomonas oryzae* pv. *oryzae* phenotypic adaptation in a semi-isolated environment. Scientific Reports. 2016;**6**:34137. DOI: 10.1038/srep34137.

[22] Doucouré H et al. Functional and genome sequence-driven characterization of tal effector gene repertoires reveals novel variants with altered specificities in closely related Malian *Xanthomonas oryzae* pv. *oryzae* strains. Frontiers in Microbiology. 2018;**9**:1657

[23] Tran TT et al. Functional analysis of African *Xanthomonas oryzae* pv. *oryzae* TALomes reveals a new susceptibility gene in bacterial leaf blight of rice. PLoS Pathogens. 2018;**14**:1-25

[24] Midha S et al. Population genomic insights into variation and evolution of *Xanthomonas oryzae* pv. *oryzae*. Scientific Reports. 2017;7:40694

[25] Song WY et al. A receptor kinase-like protein encoded by the rice disease resistance gene, Xa21. Science. 1995;**270**(5243):1804-1806

[26] Sun X et al. Xa26, a gene conferring resistance to *Xanthomonas oryzae* pv. *oryzae* in rice, encodes an LRR receptor kinase-like protein. The Plant Journal. 2004;**37**(4):517-527

[27] Pruitt RN et al. The rice immune receptor XA21 recognizes a tyrosine-sulfated protein from a Gram-negative bacterium. Science Advances. 2015;**1**(6):e1500245

[28] Gómez-Gómez L, Boller T. Flagellin perception: A paradigm for innate immunity. Trends in Plant Science. 2002;**7**(6):251-256

[29] Zipfel C et al. Perception of the bacterial PAMP EF-Tu by the receptor EFR restricts Agrobacterium-mediated transformation. Cell. 2006;**125**(4):749-760

[30] Shiu S-H, Bleecker AB. Plant receptor-like kinase gene family: Diversity, function, and signaling. Science's STKE. 2001;**2001**(113):re22

[31] Yoshimura S et al. Expression of Xa1, a bacterial blight-resistance gene in rice, is induced by bacterial inoculation. Proceedings of the National Academy of Sciences. 1998;**95**(4):1663-1668

[32] Ji Z et al. Interfering TAL effectors of *Xanthomonas oryzae* neutralize R-gene-mediated plant disease resistance. Nature Communications. 2016;7:13435. DOI: 10.1038/ncomms13435

[33] Triplett LR et al. A resistance locus in the American heirloom rice variety Carolina Gold Select is triggered by TAL effectors with diverse predicted targets and is effective against African strains of *Xanthomonas oryzae* pv. *oryzicola*. The Plant Journal. 2016;**87**(5):472-483

[34] Read AC et al. Suppression of Xo1-mediated disease resistance in rice by a truncated, non-dna-binding TAL effector of *Xanthomonas oryzae*. Frontiers in Plant Science. 2016;**7**(1516):eCollection

[35] Zhang J, Yin Z, White F. TAL effectors and the executor geneR genes. Frontiers in Plant Science. 2015;**6**:641

[36] Gu K et al. High-resolution genetic mapping of Xa27 (t), a new bacterial

blight resistance gene in rice, *Oryza sativa* L. Theoretical and Applied Genetics. 2004;**108**(5):800-807

[37] Gu K et al. geneR gene expression induced by a type-III effector triggers disease resistance in rice. Nature. 2005;**435**(7045):1122

[38] Tian D et al. The rice TAL effector-dependent resistance protein XA10 triggers cell death and calcium depletion in the endoplasmic reticulum. The Plant Cell. 2014. DOI: 10.1105/tpc. 113.119255

[39] Wang C et al. XA23 is an executor R protein and confers broad-spectrum disease resistance in rice. Molecular Plant. 2015;**8**(2):290-302

[40] Hu K et al. Improvement of multiple agronomic traits by a disease resistance gene via cell wall reinforcement. Nature Plants. 2017;**3**:17009

[41] Chen LQ et al. Sugar transporters for intercellular exchange and nutrition of pathogens. Nature. 2010;**468**(7323):527-532

[42] Yang B, Sugio A, White FF. Os8N3 is a host disease-susceptibility gene for bacterial blight of rice. Proceedings of the National Academy of Sciences. 2006;**103**(27):10503-10508

[43] Antony G et al. Rice xa13 recessive resistance to bacterial blight is defeated by induction of the disease susceptibility gene Os-11N3. Plant Cell. 2010;**22**(11):3864-3876

[44] Zhou J et al. Gene targeting by the TAL effector PthXo2 reveals cryptic resistance gene for bacterial blight of rice. The Plant Journal. 2015;**82**(4):632-643

[45] Streubel J et al. Five phylogenetically close rice SWEET genes confer TAL effector-mediated susceptibility to *Xanthomonas oryzae* pv. *oryzae*. The New Phytologist. 2013;**200**(3):808-819

[46] Yu Y et al. Colonization of rice leaf blades by an African strain of *Xanthomonas oryzae* pv. *oryzae* depends on a new TAL effector that induces the rice nodulin-3 Os11N3 gene. Molecular Plant-Microbe Interactions. 2011;**24**(9):1102-1113

[47] Chu Z et al. Promoter mutations of an essential gene for pollen development result in disease resistance in rice. Genes and Development. 2000;**20**(10):1250-1255

[48] Gust AA, Brunner F, Nurnberger T. Biotechnological concepts for improving plant innate immunity. Current Opinion in Biotechnology. 2010;**21**(2):204-210

[49] Iyer-Pascuzzi AS, McCouch SR. Recessive resistance genes and the *Oryza sativa-Xanthomonas oryzae* pv. *oryzae* pathosystem. Molecular Plant-Microbe Interactions. 2007;**20**(7):731-739

[50] Hutin M et al. A knowledge-based molecular screen uncovers a broad-spectrum Os SWEET 14 resistance allele to bacterial blight from wild rice. The Plant Journal. 2015;**84**(4):694-703

[51] Huang S et al. The broadly effective recessive resistance gene xa5 of rice is a virulence effector-dependent quantitative trait for bacterial blight. The Plant Journal. 2016;**86**(2):186-194

[52] Liu Q et al. A paralog of the MtN3/salivafamily recessively confers race-specific resistance to *Xanthomonas oryzae* inrice. Plant, Cell and Environment. 2011;**34**(11):1958-1969

[53] Zaka A et al. Natural variations in the promoter of OsSWEET13 and OsSWEET14 expand the range of resistance against *Xanthomonas oryzae* pv. *oryzae*. PLoS ONE. 2018;**13**(9):e0203711

[54] Iyer AS, McCouch SR. The rice bacterial blight resistance gene xa5 encodes anovel form of disease resistance. Molecular Plant-Microbe Interactions. 2004;**17**(12):1348-1354

[55] Jiang G-H et al. Testifying the rice bacterial blight resistance gene xa5 by genetic complementation and further analyzing xa5 (Xa5) in comparison with its homolog TFIIAγ1. Molecular genetics and Genomics. 2006;**275**(4):354-366

[56] Yuan M et al. A host basal transcription factor is a key component for infection of rice by TALE-carrying bacteria. eLife. 2016;**5**

[57] Carpenter SCD et al. A strain of an emerging Indian *Xanthomonas oryzae* pv. *oryzae* pathotype defeats the rice bacterial blight resistance gene xa13 without inducing a clade III SWEET gene and is nearly identical to a recent Thai isolate. Frontiers in Microbiology. 2018;**9**:2703

[58] Bai J et al. *Xanthomonas oryzae* pv. *oryzae* avirulence genes contribute differently and specifically to pathogen aggressiveness. Molecular Plant-Microbe Interactions. 2000;**13**(12):1322-1329

[59] Yang B, White FF. Diverse members of the AvrBs3/PthA family of type III effectors are major virulence determinants in bacterial blight disease of rice. Molecular Plant-Microbe Interactions. 2004;**17**(11):1192-1200

[60] Yang Y, Gabriel DW. Intragenic recombination of a single plant pathogen gene provides a mechanism for the evolution of new host specificities. Journal of Bacteriology. 1995;**177**(17):4963-4968

[61] Yang B, Sugio A, White FF. Avoidance of host recognition by alterations in the repetitive and C-terminal regions of AvrXa7, a type III effector of *Xanthomonas oryzae* pv. *oryzae*. Molecular Plant-Microbe Interactions. 2005;**18**(2):142-149

[62] Cruz CMV et al. Predicting durability of a disease resistance gene based on an assessment of the fitness loss and epidemiological consequences of avirulence gene mutation. Proceedings of the National Academy of Sciences. 2000;**97**(25):13500-13505

[63] Cao J et al. Dominant and recessive major geneR genes lead to different types of host cell death during resistance to *Xanthomonas oryzae* in rice. Frontiers in Plant Science. 2018;**9**:1711

[64] Yang B et al. The virulence factor AvrXa7 of *Xanthomonas oryzae* pv. *oryzae* is a type III secretion pathway-dependent nuclear-localized double-stranded DNA-binding protein. Proceedings of the National Academy of Sciences. 2000;**97**(17):9807-9812

[65] Song C, Yang B. Mutagenesis of 18 type III effectors reveals virulence function of XopZ(PXO99) in *Xanthomonas oryzae* pv. *oryzae*. Molecular Plant-Microbe Interactions. 2010;**23**(7):893-902

[66] Zhu W, Magbanua MM, White FF. Identification of two novel hrp-associated genes in the hrp gene cluster of *Xanthomonas oryzae* pv. *oryzae*. Journal of Bacteriology. 2000;**182**(7):1844-1853

[67] Sugio A et al. Two type III effector genes of *Xanthomonas oryzae* pv. *oryzae* control the induction of the host genes OsTFIIAγ1 and OsTFX1 during bacterial blight of rice. Proceedings of the National Academy of Sciences. 2007;**104**(25):10720-10725

[68] Hummel AW, Doyle EL, Bogdanove AJ. Addition of transcription activator-like effector binding sites to a pathogen strain-specific rice bacterial blight resistance gene makes it effective against

additional strains and against bacterial leaf streak. The New Phytologist. 2012;**195**(4):883-893

[69] Zeng X et al. Genetic engineering of the Xa10 promoter for broad-spectrum and durable resistance to *Xanthomonas oryzae* pv. *oryzae*. Plant Biotechnology Journal. 2015;**13**(7):993-1001

[70] Römer P, Recht S, Lahaye T. A single plant resistance gene promoter engineered to recognize multiple TAL effectors from disparate pathogens. Proceedings of the National Academy of Sciences. 2009;**106**(48):20526-20531

[71] Boch J et al. Breaking the code of DNA binding specificity of TAL-type III effectors. Science. 2009;**326**(5959):1509-1512

[72] Moscou MJ, Bogdanove AJ. A simple cipher governs DNA recognition by TAL effectors. Science. 2009;**326**(5959):1501

[73] Romer P et al. Promoter elements of rice susceptibility genes are bound and activated by specific TAL effectors from the bacterial blight pathogen, *Xanthomonas oryzae* pv. *oryzae*. The New Phytologist. 2010;**187**(4):1048-1057

[74] Morbitzer R et al. Regulation of selected genome loci using de novo-engineered transcription activator-like effector (TALE)-type transcription factors. Proceedings of the National Academy of Sciences of the United States of America. 2010;**107**(50):21617-21622

[75] Zhang F et al. Programmable sequence-specific transcriptional regulation of mammalian genome using designer TAL effectors. Nature Biotechnology. 2011;**29**(2):149

[76] Li T et al. Designer TAL effectors induce disease susceptibility and resistance to *Xanthomonas oryzae* pv. *oryzae* in rice. Molecular Plant. 2013;**6**(3):781-789

[77] Li T et al. Modularly assembled designer TAL effector nucleases for targeted gene knockout and gene replacement in eukaryotes. Nucleic Acids Research. 2011;**39**(14):6315-6325

[78] Christian M et al. Targeting DNA double-strand breaks with TAL effector nucleases. Genetics. 2010;**186**(2):757-761

[79] Cermak T et al. Efficient design and assembly of custom TALEN and other TAL effector-based constructs for DNA targeting. Nucleic Acids Research. 2011;**39**(12):e82

[80] Li T et al. TAL nucleases (TALNs): Hybrid proteins composed of TAL effectors and FokI DNA-cleavage domain. Nucleic Acids Research. 2011;**39**(1):359-372

[81] Chen K, Gao C. Targeted genome modification technologies and their applications in crop improvements. Plant Cell Reports. 2014;**33**(4):575-583

[82] Wiedenheft B, Sternberg SH, Doudna JA. RNA-guided genetic silencing systems in bacteria and archaea. Nature. 2012;**482**(7385):331

[83] Jinek M et al. A programmable dual-RNA-guided DNA endonuclease in adaptive bacterial immunity. Science. 2012;**337**(6096):816-821

[84] Hsu PD, Lander ES, Zhang F. Development and applications of CRISPR-Cas9 for genome engineering. Cell. 2014;**157**(6):1262-1278

[85] Blanvillain-Baufumé S et al. Targeted promoter editing for rice resistance to *Xanthomonas oryzae* pv. *oryzae* reveals differential activities for SWEET14-inducing TAL effectors. Plant Biotechnology Journal. 2017;**15**(3):306-317

[86] Li T et al. High-efficiency TALEN-based gene editing produces

disease-resistant rice. Nature Biotechnology. 2012;**30**(5):390-392

[87] Hutin et al. Ectopic activation of the rice NLR heteropair RGA4/RGA5 confers resistance to bacterial blight and bacterial leaf streak diseases. The Plant Journal. 2016;**88**(1):43-55

[88] Tian D, Yin Z. Constitutive heterologous expression of avrXa27 in rice containing the *R* gene Xa27 confers enhanced resistance to compatible *Xanthomonas oryzae* strains. Molecular Plant Pathology. 2009;**10**(1):29-39

[89] Schwessinger B et al. Correction: Transgenic expression of the dicotyledonous pattern recognition receptor EFR in rice leads to ligand-dependent activation of defense responses. PLoS Pathogens. 2015;**11**(4):e1004872

[90] Lu F et al. Enhancement of innate immune system in monocot rice by transferring the dicotyledonous elongation factor Tu receptor EFR. Journal of Integrative Plant Biology. 2015;**57**(7):641-652

[91] Thomas NC et al. The rice XA21 ectodomain fused to the Arabidopsis EFR cytoplasmic domain confers resistance to *Xanthomonas oryzae* pv. *oryzae*. PeerJ. 2018;**6**:e4456

Section 3

Modern Cultural Practices

Chapter 9

Protecting Rice Grains from Arsenic Toxicity through Cultural Management: Bangladesh Perspective

Abdul Aziz

Abstract

In 1997, arsenicosis was reported as a result of ingesting arsenic-containing rice grown in arsenic (As)-rich soil, irrigated with high As water from shallow tube wells (STW) and deep tube wells (DTW) in Bangladesh. Of the 4 million ha irrigated fields, 60% were under STW and 15% under DTW waters; almost all were arsenic contaminated in varying quantities since they were used. In the present study, it was determined that irrigation from STW water having 500 μg As/l produced rice grains with 2.56 mg As/kg in a field with initial 3.21 mg/kg soil, leaving 8.27 mg/kg soil compared to pond water irrigation where only roots absorbed 0.105 ± 0.069 mg As/kg leaving ≤2.6 mg/kg soil. About 2.5 mg As/kg soil may be considered a safe level for arsenic-free rice cultivation. Bio-mitigation of the STW water using duckweed (DW) (*Spirodela polyrhiza*) was expensive and disposal in various ways of As-loaded DW produced was hazardous returning arsenic to ecosystems. Alternative to the groundwater (GW), surface water can be made available by constructing rubber dams and converting rivers into surface water reservoirs to overcome the arsenic toxicity and protecting rice and other grains, integrating aquaculture of the DW and *Azolla pinnata* var. *pinnata* for fish and poultry feeds. Permanent solution could be achieved executing "Delta Plan 2100" saying "No to groundwater use for irrigation, let the Arsenic stay in the underground".

Keywords: arsenic toxicity, rice toxicity, groundwater arsenic, soil arsenic, bio-mitigation, *Spirodela polyrhiza*, river reservoirs, rubber dam, Bangladesh

1. Introduction

In Bangladesh, As was discovered in 1993, while doctors and health personnel were dealing with health effects of its contamination in drinking water. Since then, As-contaminated groundwater was found in 44 districts out of which arsenicosis patients were detected in 26 districts, 7 of which were highly affected, and out of 64 districts [1] (**Figure 1**), some districts as a catastrophe affecting human health [1, 2]. Southern districts in particular contained >300 μg As/l in GW, and more than 20% tube wells contain more than 100 μg As/l that are used for irrigation and drinking [3]. The metalloid at low concentration (10–50 μg As/l) in a sandy soil may be more phytotoxic (i.e., available) than much higher levels (200–500 μg As/l) in a heavier clay soil [4]. Arsenic is found everywhere in traces, i.e., in the air, in the ocean and freshwaters (some drinking water supplies), in soil, etc., polluting the environment and causing arsenicosis (melanosis, keratosis, gangrene, chronic ulcer, skin cancer, etc.) in human [1, 5]. Studies also confirmed that a substantial amount of this heavy metal is absorbed by plants [6–9]. The question is how as appeared heavily in soils, drinking, and irrigation waters of Bangladesh? In the 1960s, 4 million hand tube wells (HTW)

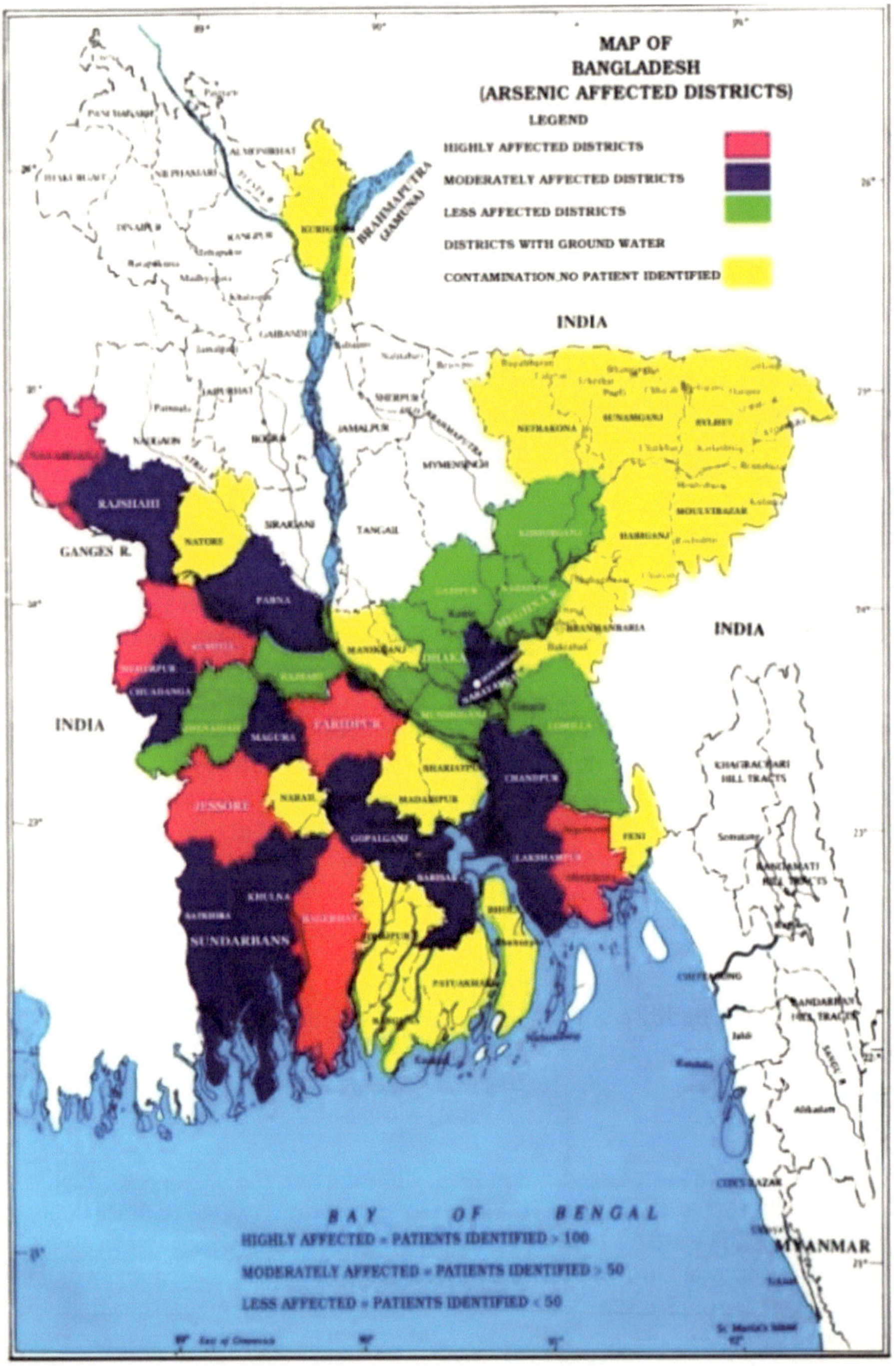

Figure 1.
Arsenic-affected areas in Bangladesh based on studies from 1993 to 1996 [1]. Round white spot (within blue) east of mid-region is the study site at Sonargaon.

were installed at a depth of 10–15 m for drinking water without checking arsenic in the aquifers [10]. In the 1970s high-yielding rice variety known as the International Rice Research Institute (IRRI) paddy var. IR8 was introduced in Bangladesh that required huge irrigation, necessitating the installation of STW at a depth of 25–30 m [personal communication Director General, Dr. Md. Shahjahan Kabir of Bangladesh Rice Research Institute, Gazipur, Bangladesh]. For decades DTW have been installed at 100–200 m depth which are also contaminated with arsenic [11]. In Jessore alone, 74 out of 85 DTW tested had an average 210 μg As/l. Of the 4 million ha irrigated fields, 25% used surface water, 60% STW, and 15% DTW waters in Bangladesh [11]. The organic As (oAs) is not but inorganic arsenic (iAs) is toxic to human [12]. Concentration of absorbed arsenic in rice grains of Bangladesh was in toxic level for human consumption [13–15]. It was reported that 1.7–2.55 mg/kg rice grains were found in areas having 15–27 mg As/kg soil [16]. It was also reported that As concentrations in rice grains from flooded soils were 10–15 times higher than aerobically grown (non-flooded) rice, and similarly the concentration of inorganic As (iAs) in the grains from the flooded treatment was 2.6–2.9 fold higher than the non-flooded treatment [15].

Rice is by far the largest food dietary source of iAs for populations not drinking water with elevated As [17]. Arsenic in rice, cowpeas, vegetable crops, peas, snap beans, and sweet corn plants antagonize the uptake of nutrients like N, P, K, S, Ca, Mn, and Z [18–20] and reduced height of rice plants, straw and grain yield, grain per panicle, and the number of field grains per panicle applying 10 mg As/kg soil or above [19–22]. Phosphate replacement by arsenate preventing ATP generation was observed by [4, 7, 23] reducing ATP-dependent N_2 fixation [9]. Human activities like smelting, mining, the use of pesticides, making glasses and ceramics, etc. are responsible for contaminating earth surface [24], which is now a global health issue [25]. However, it has been estimated that about 30 metric ton of arsenic is borne by the biomass of the earth and is assumed to be the 12th most abundant element in the biosphere [26].

Arsenic removal efficiency varies with many conditions, like site-specific chemicals and geographic and economic conditions [27]. There have not been any significant and innovative improvements in the methods for removing arsenic from HTW water in a decade of research for drinking [28]. However, no attempt has so far been taken to remove arsenic from large volume of the contaminated STW waters for using in irrigation. In the present study, an attempt was made to use efficient arsenic-absorbing floating plant *S. polyrhiza* to remove toxic arsenic from STW water in situ for protecting rice grains through cultural management.

2. Materials and methods

Field studies were carried out at Nilkanda Union under P. S. Sonargaon, District Narayanganj, lying between 23°30′ and 23°46′ N and 90°31′ and 90°41′ E (**Figure 1**) [29]. A good number of submerged and exposed soil (rice, wheat, and vegetable fields) and water (STW, HTW, and pond) samples were collected in four replications at a gap of 0.5–1.0 km in the winter and spring seasons and analyzed to determine the amount of As absorbed by different crops and its presence in soils and water sources. Water samples were collected from DTW and STW one from each at village Kachua, under P.S. Kachua, district Chandpur, about 100 km southeast of Sonargaon.

2.1 Selection of plants and cultivation for bioremediation of arsenic

Six strains under three species of *Azolla* (*A. caroliniana* Dh 103; *A. filiculoides* Dh 104; *A. pinnata* var. *pinnata* Dh 111; *A. pinnata* var. *pinnata* Dh 112; *A. pinnata* var. pinnata Dh 113; and *A. filiculoides* Dh 115) and two species of *Spirodela* (*S. polyrhiza* Dh

116 and *S. punctata* Dh 117, Lemnaceae) were considered to select efficient As absorbing floating plants [9] by growing them in the growth room under controlled conditions: continuous light flux of 120 μE/m^2/s from daylight fluorescent tubes and 70–90% relative humidity at 30 ± 1°C. Liquid inorganic nutrient medium Chu 10D-N [30], a modified version of [31], with double the strength of P (3.56 mg/l) and K (9.45 mg/l) for *Azolla* strains, and IRRI medium with NH_4NO_3 as nitrogen source following [32] for species of *Spirodela* were grown in conical flasks of 150 ml capacity containing 50 ml medium adding arsenic trioxide as inorganic As (iAs) source before autoclaving. *S. polyrhiza* and *S. punctata* were grown at much higher arsenic concentrations as the plant is often found to grow in polluted waters and stressed environment [33]. The plants were grown at pH 6.00 [34, 35]. All plant biomasses after 3 days growth were oven dried at 90°C for 24 hour for determining growth and As absorbed [9].

2.2 Determination of nitrogenase activity

Nitrogenase activity in the presence of arsenic was determined by acetylene reduction assay (ARA) technique of [36] and details of the experiment given in [9].

2.3 Chemical analysis

Home-yard ponds and HTW waters were collected in iodized bottles and stored in ice box immediately and then in the laboratory deep freeze. Water quality of the pond and HTW waters is given in **Figure 2**. Whole plants of *Azolla* and *Spirodela*, roots, shoots, and grain samples of rice plant were collected separately and dried at 70°C for 3 days, grounded, and kept in poly-bags till analysis. Arsenic was determined by atomic absorption spectrometer (Shimadzu AA-680/G V-3) after digestion with nitric and perchloric acids at 5:1 ratio in a closed system. Arsenic concentration in brown-rice grains was cross-checked by neutron activation analysis

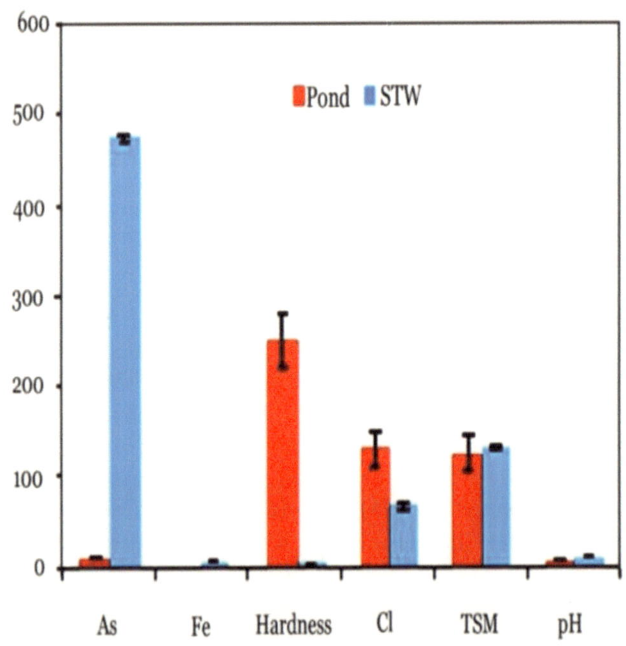

Figure 2.
Water quality of pond (control) and STW (installed 3 years ago) used for growing BR-28 rice in fields. Unit for tAs is μg/l and for others as mg/l. After [43]. n = 5; vertical bars are standard deviations.

(NAA) using 3 MW Research Reactor at Atomic Energy Research Establishment, Saver, Dhaka. PO_4-P in the culture medium was determined following [37].

Soil samples were collected from a depth of 0–5 cm in five replications from each plot on the basis of composite sampling method in [38]. Determinations of various fractions were carried out as described: EC by EC meter in [39]; pH by a microprocessor pH-meter, soil to water ratio was 1:2.5; organic carbon by wet oxidation method; organic matter by multiplying the percentage of organic carbon with conventional Van Bemmelen's factor of 1.74 in [40]; textural classes and particle size distribution by hydrometer method as described in [41, 42], respectively; and As by AAS (Perkin Elmer M-3110, USA) following low temperature sample digestion with HNO_3-HCl mixture at a ratio of 1:3 in volumetric flasks with reflex condenser under a closed system.

2.4 Cultivation of *S. polyrhiza* in a production pond

The DW was cultivated in a 400 m^2 pond adding urea nitrogen, triple super phosphate, and muriate of potash in a solution at the rate of 40, 20, and 10 kg/ha/day, respectively [29]. The fertilizer solution was mixed with the pond water every day just after 10% harvest before noon time. The production was 1 ton/ha/day and 1 kg DW covers about 1 m^2 water.

2.5 Bioremediation of arsenic-contaminated HTW water

Mini-scale bioremediation was carried out in 50 L capacity RCC (cemented) tubs lined with polythene [29]. The tubs were set in the open area of a house having sufficient natural light. Arsenic-contaminated HTW (installed 25 years ago) having 475.5 ± 10.6 μg/l and 6.30 mg/l iron was used. Fresh DW from the cultivation pond was spread onto the tub water to form a complete cover. The amount per tub added was weighed, and the same amount was added after every 24 hours for 6 days. 100 ml surface water, 20 g fresh DW, and sediment from each tub bottom were collected for analysis of As in the samples.

Large-scale bioremediation experiment was carried out in about 1 meter deep, 350 m^2 pond situated near to the experimental rice field [29]. It was filled with arsenic (495 ± 10 μg As/l was considered as 500 μg As/l)-contaminated STW water which appeared reddish brown in color. The STW water had 8.85 mg/l iron. About 350 kg DW was spread over the stored STW water in the pond and divided into eight blocks by bamboos for keeping the plants equally spread for As absorption. After every 2 days, 1.0 kg fresh DW, 100 ml water, and 50 g sediment were collected from each block for chemical analysis and for using as feed for broiler and Bengal goat. All the floating DW was then harvested (kept in a shallow pit for decomposition and used in biogas production). Fresh DW was again spread completely covering the water surface.

2.6 Cultivation of rice

Effects of arsenic on the arsenic toxicity of winter-rice grain of variety BR28 were studied in a farmer's field having silt clay containing sand 4.31%, silt 45.56%, clay 50.13%, pH 6.35, EC 200 μS/cm, iron 0.87%, arsenic 3.21 mg/kg soil, and organic matter 1.85%, in Silmondi and Narailbag soil series at Meghna floodplain [29]. Research plots were 19 × 19 m in five replications arranged in complete randomized block design. NPK fertilizers from urea, TSP, and MP, respectively, were applied at 45, 15, and 40 kg/ha, respectively, to all plots at the time of land preparation other than N. The N was applied at 25 kg/ha 15 days after transplanting (DAT) and 20 kg/ha before flowering. "Boro" rice var. BR28 seedlings of 10 days old were transplanted and irrigated once in a week with pond water as control and 500 μg As/l contaminated STW water that dries up by the end of that week and watered again for the next one week [29].

3. Results and discussion

3.1 Water and soil chemistry

Prevalence of total arsenic (As) in soil and water sources collected from areas at a gap of 0.5–1.0 km in the winter and spring seasons at Sonargaon is given in **Table 1**. Arsenic in pond water was due to seepage and water flow from the household use of As-contaminated HTW. The concentration of iron had direct correlation to the amount of arsenic present. Rice field soils in the areas had highest arsenic ranging from 5.83 to 8.01 mg/kg (due to weekly irrigation about nine times with STW water), followed by wheat (due to broadcast seeding and standing irrigation after 20, 60, and 80 days (before flowering)) and vegetable fields (due to non-standing irrigation two times during dry period). Water from the HTW was found to have 131 ± 0.1 μg As/l (10 years old) to 475.5 ± 10.6 μg As/l (25 years old). Similarly, 1-year-old STW water was found to have 92.2 ± 1.5 μg As/l, while the 3-year-old one had 495.0 ± 10 μg As/l indicating that the older the tube well, the more is the groundwater arsenic indicating tube wells having less than 100 μg As/l in the early 2000s most likely have now increased to several hundred or more.

The water of 1-year-old DTW was with very little or no arsenic, while STW had 150 ± 3.0 μg As/l and has been found to be moderately affected [1] at village Kachua in Meghna floodplain. In Ganges floodplain, there are reports of an average 210 μg As/l in DTW water in many areas of Jessore district and in highly affected areas (**Figure 1**). The most alarming point is that each and every crop had arsenic toxicity for irrigating with GW (**Table 1**). To get a good yield, irrigation is a must and alternative source(s) need to be found out.

Quality of pond and STW waters used in irrigating rice cultivation experiments is shown in **Figure 2**.

3.2 Selection of arsenic removing plants in the laboratory

Six strains of *Azolla* under three species were treated with arsenic trioxide in the laboratory to determine their ability to grow and absorb As, PO_4-P uptake, and

Waters (μg/l for As, mg/l for iron)								Soils (mg/kg)		
DTW		STW		HTW		Pond				
As	Iron	As	Iron	As	Iron	As	Iron	Rice	Wheat	Vege.
3.40 ± 0.90*	0.95*	495 ± 10.00	8.85	250.5 ± 2.08	7.40	9.66 ± 1.16	0.83	7.65	4.42	3.15
		(3 years old)		(15 years old)		9.13 ± 0.31	0. 70	6.39	4.23	2.53
		313 ± 4.00	8.15	131.0 ± 0.10	6.10	3.20 ± 0.20	0. 44	7.88	3.15	1.96
		(2 years old)		(10 years old)		3.11 ± 0.17	0.30	8.01	4.09	4.21
		92.2 ± 1.50	8.10	152.0 ± 3.5	6.30			5.83		3.65
		(1 year old)		(10 years old)				7.48		3.68
		150 ± 3.00*	6.10*	475.5 ± 10.6	7.60					4.02
		(3 years old)		(25 years old)						2.66

**Indicate values of two tube wells from village Kachua P.S. Kachua (moderately affected area).*

Table 1.
Prevalence of arsenic in soils and waters at Sonargaon.

nitrogenase activity (**Figure 3a–d**) [9]. *A. pinnata* var. *pinnata* Dh 111, Dh 112, and Dh 113 strains grew well at 100 μg iAs/l. The growth of strains Dh 112 and Dh113 was identical even at 400 μg iAs/l in 3 days (**Figure 3b**). Only *A. pinnata* var. *pinnata* Dh 113 appeared to be suitable for absorption of iAs from 100 μg iAs/l [9]. This was due to the substantial absorption of PO_4-P at 100 μg iAs/l and mild absorption in higher concentrations (**Figure 3c**). *A. pinnata* var. *pinnata* Dh 113 absorbed about 0.0066% arsenic of d. wt. from 100 to 200 μg/l As in 3 days (**Figure 3b**) [9]. Only *A. pinnata* var. *pinnata* Dh 113 appeared to be suitable for absorption of As from 100 μg/l As [43]. The presence of higher amount of PO_4-P in the iAs containing medium (**Figure 3c**) than the control (without iAs) suggests that after three days growth phosphate uptake was limited by plants and replaced by arsenate or competitively absorbed limiting ATP formation, indicated by decreased ATP-dependent nitrogenase activity, which was nil at 400 μg As/l (**Figure 3d**) [9].

S. polyrhiza Dh 116 and *S. punctata* Dh 116 were treated with arsenic trioxide in the laboratory to determine their ability to grow and absorb As and PO_4-P (**Figure 4a–c**) [9]. The highest accumulation of 0.0351% iAs on dry wt. basis (**Figure 4b**) was observed in *S. polyrhiza* from 1000 μg iAs/l, and this was due to substantial absorption of PO_4-P from medium containing 1000 μg iAs/l (**Figure 4c**).

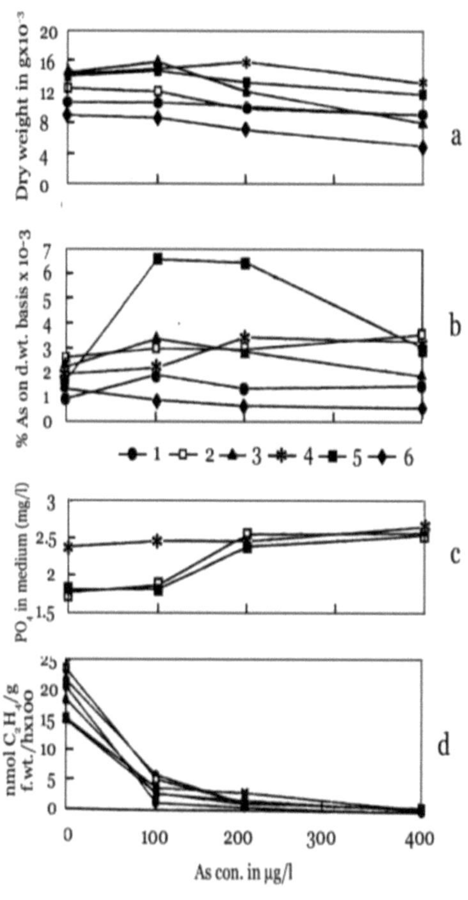

Figure 3.
(a–d) Effects of arsenic trioxide in batch culture under controlled environments measured after 3 days of inoculation: (a) on growth, (b) % accumulation of As, (c) absorption of PO_4-P, and (d) nitrogenase activity by Azolla species/strains—(1) A. caroliniana Dh103, (2) A. filiculoides Dh104, (3) A. pinnata var. pinnata Dh111, (4) A. pinnata var. pinnata Dh112, (5) A. pinnata var. pinnata Dh113, and (6) A. filiculoides Dh115. After [9].

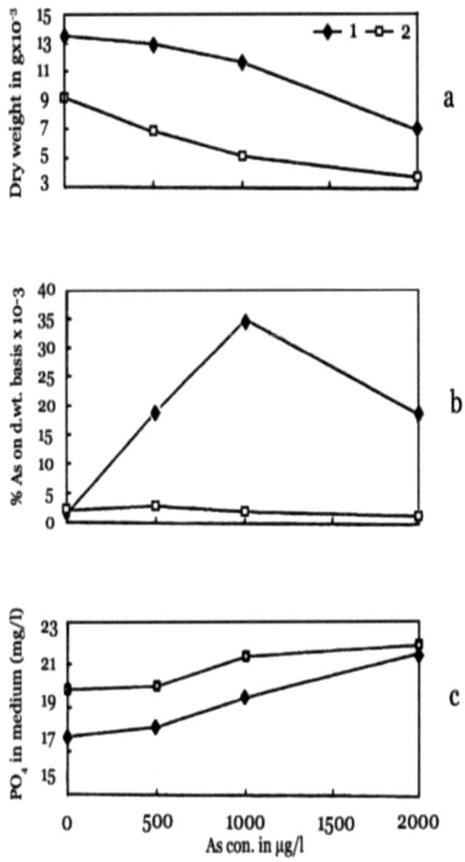

Figure 4.
(a–c) Effects of arsenic trioxide in batch culture under controlled environments measured after 3 days of inoculation: (a) on growth, (b) % accumulation of As, and (c) absorption of PO_4-P by Spirodela—(1) S. polyrhiza Dh116 and (2) S. punctata Dh117. After [9].

Of the eight floating plants tested, *S. polyrhiza* Dh 116 showed the highest absorption capacity to be much higher than *A. pinnata* var. *pinnata* Dh 113 [9].

3.3 Bio-mitigation of As from contaminated STW water using *S. polyrhiza*

3.3.1 Removing As in a mini-scale with the water in cemented tubs

In the laboratory *S. polyrhiza* showed the highest accumulation (0.0351% As on dry wt. basis) among floating plants from 1000 µg/l in 3 days (**Figure 4b**) [9]. Therefore, outdoor experiments were carried out at Sonargaon growing the DW in RCC tubs in 475.5 ± 10.6 µg As/l water and 6.30 mg Fe/l (**Figure 5a–c**). The DW absorbed about 295 mg As/kg d. wt. after 24-hour treatment (**Figure 5a**). A substantial amount of As was coagulated with iron (synergistic reaction) from 24 hours to 6 days giving similar curve like absorption of As by DW (**Figure 5b** and **c**).

3.3.2 Removing As in a large-scale keeping the water in a pond

The *S. polyrhiza* Dh 116 absorbed about 295 mg/kg d. wt. after 24 hours in tub experiment and thus could be a good candidate for mitigating As from the contaminated STW water in a large scale (**Figure 6a–c**). The 350 m^2 pond contains 350 m^3

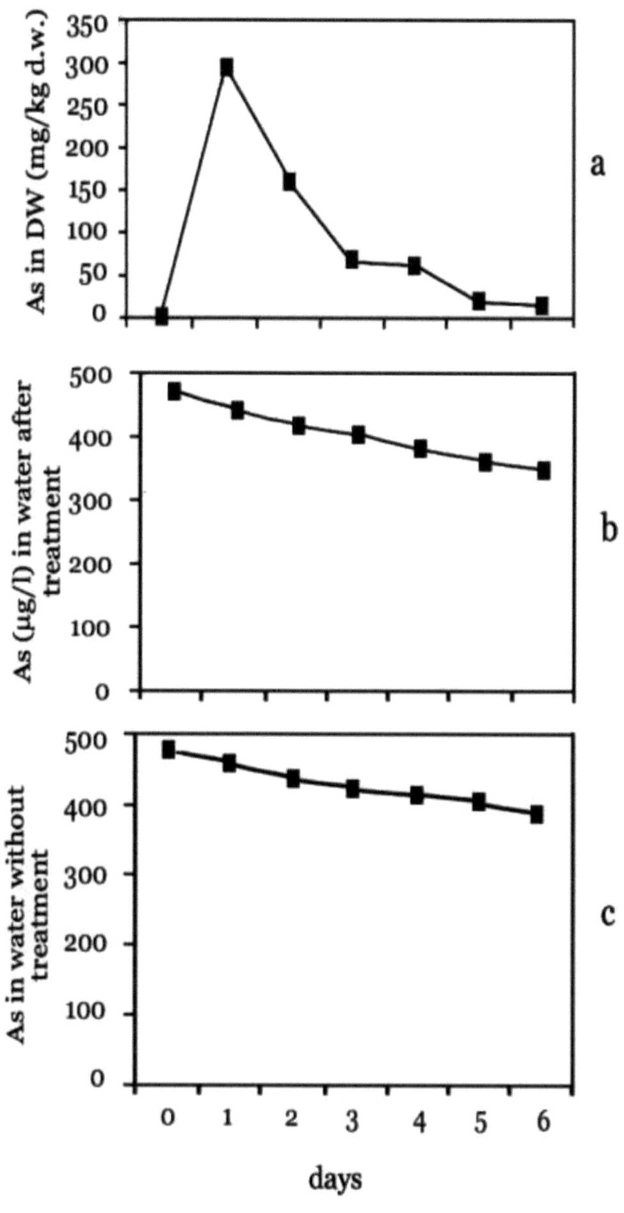

Figure 5.
(a–c) Changes in concentration of 475.5 ± 10.6 µg As/l HTW water after treatment with S. polyrhiza dh 116 in cemented (RCC) tubs.

or 350,000 l water. One liter STW water contains 500 µg As/l. Three hundred fifty thousand liters of water would contain 175 g As from which 157.5 g is to be removed to get 50 µg As/l acceptable level of irrigation water. The DW removed 325 mg As/kg dry DW (**Figure 7a**) with similar iron absorption curve in 2 days (**Figure 7b**). Therefore, 350 kg fresh DW was equal to 20 kg dry DW (5% basis) which could remove 6500 mg or 6.5 g As, and thus to remove 157.5 g arsenic 24-hour treatment means 8.4 ton fresh DW would be needed in 48 days. In each 24 hour, arsenic loaded roots and plant debris, and As and Fe coagulates (synergistic) deposited 63.7 mg As on the pond bottom (**Figure 7d** and **e**), estimated to be about 1.529 g As after treatment. The bioremediation technique is time-consuming and expensive, requiring two ponds and over eight tons As-loaded DW waste.

Figure 6.
(a–c) Bioremediation process of arsenic contaminated GW using S. polyrhiza. (a) 500 μg/l As STW water immediately after storage in the pond appeared grayish red colored, (b) a complete cover of the DW on the water, and (c) half of the DW cover removed after 8 days of treatment showing relatively clear water. Inset in "c" shows dorsal and ventral surfaces of the plant.

3.4 Effects of As-contaminated irrigation water on rice cultivation

"Boro" rice BR28 was grown in 500 μg As/l STW water (Section 2.6).

3.4.1 Arsenic in rice-cultivated soil

Arsenic concentration in soils of the experimental plots during rice cultivation is shown in **Figure 8**. Initial soil As was 3.21 mg/kg, increased to 7.51 mg/kg due to low uptake of As during seedling adaptation stage for 7 days irrigation with 500 μg As/l and 8.85 mg Fe/l STW water. During March rapid vegetative growth results in rapid uptake of As, reducing it to 5.50 mg As/kg soil. Further absorption

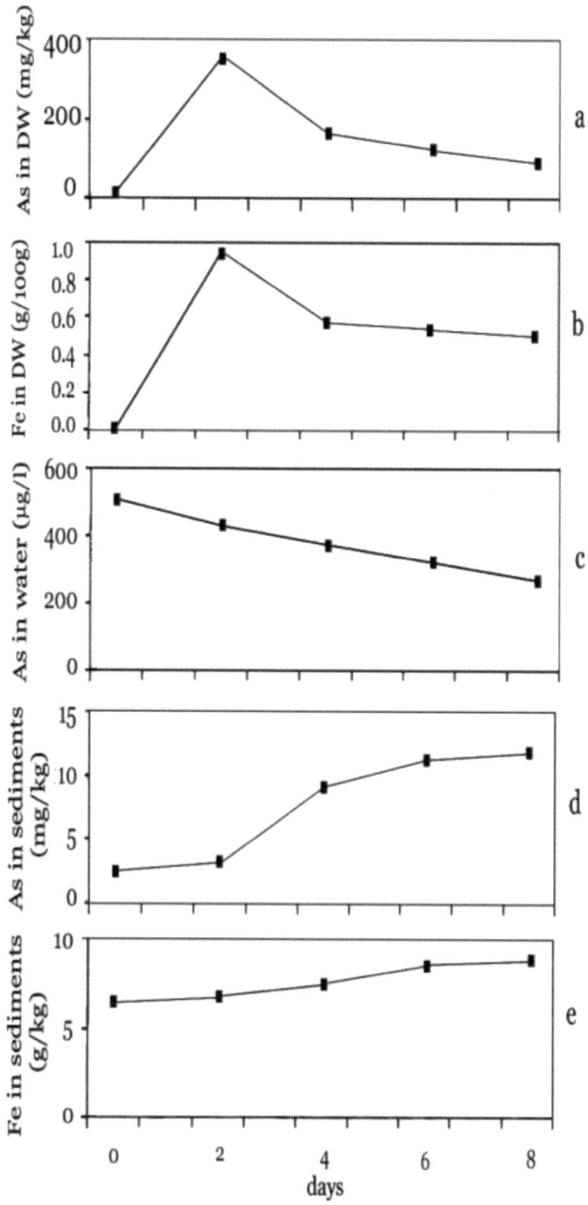

Figure 7.
(a–e) Changes in As (500 µg As/l) and Fe (8.85 mg/l iron) after treating with S. polyrhiza in the treatment pond and natural coagulation.

during panicle initiation, grain formation, and maturation in April reduced As to 4.90 mg/kg soil. Irrigation was stopped after a week creating non-flooded condition when As solubility decreased [15] simultaneously with low or no nutrient uptake by mature plants, increasing As concentration to 8.27 ± 1.35 mg/kg soil during harvest (**Figure 8**). There are reports on the increased soil As for using contaminated GW in rice fields year after year to 1.0 and 1.1 mg/kg soil in Bangladesh [16] and West Bengal [44], respectively.

In irrigated pond water (**Figure 8**), As concentration insignificantly decreases till the day of harvesting, ranging from 1.91 ± 0.26 to 2.6 ± 0.80 mg/kg soil from initial 3.21 mg/kg. Only roots absorbed 0.075–0.156 mg As/kg in the presence of a good amount of As for the growing period (**Figures 8** and **9**). It appears that As absorption in pond water

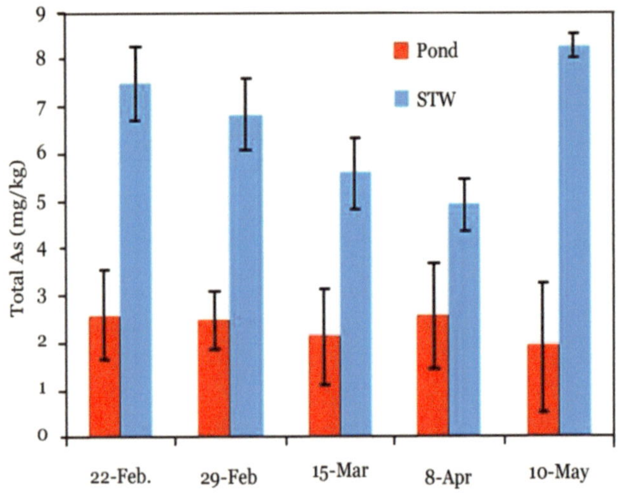

Figure 8.
Arsenic concentration in soils of the experimental plots during rice cultivation transplanted on 15 Feb 2004, irrigated with pond and shallow tube well waters. Initial soil arsenic was 3.21 mg/kg; contaminated pond water had only 9.50 ± 0.50 µg/l, while STW water had 500 µg As/l. Rice crop harvested on 10 may. After [43]. n = 5; vertical bars, standard deviation.

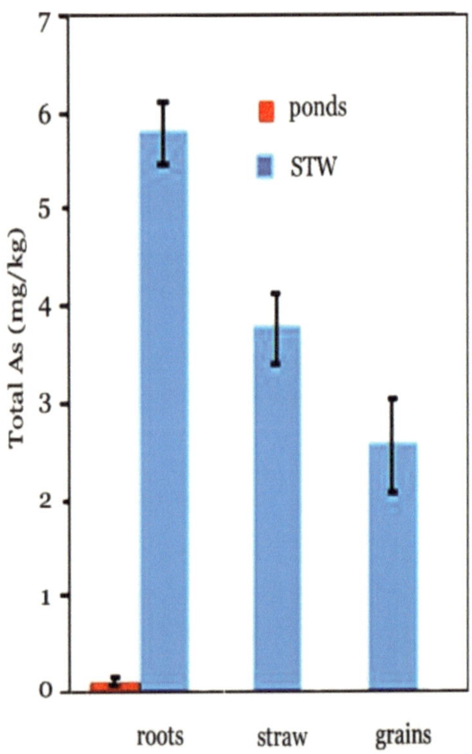

Figure 9.
Arsenic concentration in roots, straw, and brown rice grains of BR 28 at the time of harvest. Initial soil arsenic was 3.21 mg/kg, while pond and shallow tube well waters contained 9.50 ± 0.50 and 476 ± 3 µg/l arsenic, respectively. After [43]. n = 5; ± and vertical bars are standard deviation.

irrigated plots is limited at 2.6 ± 0.80 mg/kg soil. Therefore, 2.5 mg/kg soil could be considered as a safe level for arsenic-free rice cultivation in Bangladesh. The permissible limit for the USA is 5 mg As/kg soil for agricultural use [45].

It has been estimated that Bangladesh clayey soil needs about 1000 l irrigation water/m^2/year (in conservative use) [11]; in other words 2500 l/kg surface water is needed for producing 1 kg rice, considering production of 4.0 ton/ha.

3.4.2 Arsenic distributions/accumulations in rice crops

Arsenic accumulation in the "Boro" rice plant parts was mostly in roots followed by shoots, brown-rice grains (**Figure 9**), and husks. The rice grain contained 2.552 ± 0.507 mg/kg As, similar to adjacent rice field (2.56 ± 0.20 mg/kg) which was determined by neutron activation analysis. The presence of 1.7–1.8 mg/kg rice grains was recorded in areas having 15–27 mg As/kg soil [13]. The high As content in straws found in the present study might affect cattle due to bioaccumulation when consumed, while roots would contribute about 5.7 mg/kg for the next crop. Using pond water as a control in the same field, only roots absorbed 0.113 ± 0.054 mg As/kg leaving mean As 2.60 mg/kg soil.

The concentration of absorbed As in rice grains was much above the permissible limit of 1.00 mg/kg [14] for human consumption. It was estimated that elevated inorganic arsenic in rice significantly contributes to dietary arsenic intake in USA [17], which was estimated to be double (80%) in Bangladesh and India, and the rest was dimethyl-arsenate (DMA). Bioavailability of iAs from rice was reported to be high [15]. It was estimated that As concentration in rice would have to be as low as 0.050 mg/kg if consumed at 200 g/d to equate to similar exposure from drinking water at 10 μg/l [18]. Sonargaon rice contain 2.552 ± 0.507 mg As/kg which would have 0.85–1.276 mg iAs per kg rice assuming about 33–50% of As according to [46], while there are reports of 22–42% iAs, and the rest was DMA [15]. This in quantity ranged from 0.561 to 1.072 mg iAs/kg for Bangladesh rice. If the lowest amount 0.561 mg iAs/kg rice grain is considered and equates to permissible limit of 0.050 mg/l water (which would be about 0.250 mg iAs/kg consumed at 200 g/day by a 60 kg adult person), then a person in the study area is consuming double the amount of permissible iAs compared to China's food standard limit of 0.15 mg iAs/kg [38]. Thus estimated high iAs present in the rice grain at Sonargaon can lead to greatly increased exposure to chronic carcinogen.

The presence of arsenic affects PO_4-P absorption from the liquid medium by plants where phosphate is replaced by arsenates and prevents ATP synthesis [6, 9, 23]. Therefore, if the phosphorus, for example, is 50% less than the required amount in a cell, it could be assumed that the required amount of ATP synthesis would not take place affecting physiology, cell division to new cell formation, etc. in any affected organism. The high total arsenic in the soil would affect all the biotic communities including biological N_2-fixing soil bacteria, as is found in *A. pinnata* var. *pinnata* where ATP-dependent nitrogenase activity was severely affected [9].

A Rickshaw (a three-wheeler, non-mechanized) puller in the present study area had melanosis (hyper-pigmentation) or reddish spots, hairs having 3.99 ± 0.21, and nails having 8.90 ± 0.29 mg As/kg identified as "first stage group" [14] drinking 250 μg/l As for 2 years which was 8.42 mg/kg body wt. The lower value (4–8 mg As/kg samples) was due to the release of arsenic by deamination, and the accumulation may be higher in other parts of a body. Piped water from DTW was supplied for 2 years by the NGO Forum in the village of Nilkanda, Sonargaon; still farmers, day laborers, and Rickshaw pullers of many villages show symptoms of arsenicosis. The reason could be retention of arsenic in their body and most likely also eating high arsenic-containing rice, wheat, and vegetables (**Table 1**). The manifestation of the symptom (melanosis) appears to take place after some years of drinking and eating arsenic-contaminated water, rice, vegetables, etc. [17] as indicated by high amount of arsenic in nails and hairs.

3.4.3 Effects of high arsenic on weeds in rice fields

Effects of high arsenic on the occurrence of weeds in the rice field soil were studied. Out of 14 species recorded, six could not grow in the experimental rice field [47]. The Importance Value Index (IVI) indicated that *Alternanthera sessilis*, *Cynodon dactylon*, *Echinochloa colonum*, *Enhydra fluctuans*, *Hedyotis corymbosa*, and *Lippia nodiflora* are very sensitive to arsenic; *Lindernia antipoda* and *Eriocaulon setaceum* were not affected at all, while the growth of *Cyperus rotundus*, *Eclipta alba*, and *Fimbristylis* sp. was enhanced in the presence of arsenic (**Table 2**). The sensitiveness appears to be due to enhanced arsenic absorption by seeds, reduced germination percentage, radical length, and biomass accumulation leading to their death, as seen in seed germination of *Glycine max* at 25 and 100 μM sodium arsenate

Name of weeds	Den.	Freq.	Abun.	Rel. den.	Rel. freq.	Rel. abun.	IVI	H	D
Alternanthera sessilis	0.2 (–)	02 (–)	1.0 (–)	1.96 (–)	3.85 (–)	4.76 (–)	10.57 (–)	3.135 (2.751)	0.865 (0.836)
Cynodon dactylon	0.6 (–)	0.2 (–)	3.0 (–)	5.88 (–)	3.85 (–)	14.28 (–)	24.01 (–)		
Cyperus exaltatus	0.6 (0.4)	0.6 (0.2)	1.0 (2.0)	5.88 (8.33)	11.54 (7.69)	4.76 (14.81)	22.18 (30.83)		
C. rotundus	— (0.2)	— (0.2)	— (1.0)	— (4.16)	— (7.69)	— (7.41)	— (19.26)		
Eclipta alba	— (0.4)	— (0.2	— (1.0)	— (8.33)	— (7.69)	— (7.41)	— (23.43)		
Echinochloa colonum	1.0 (–)	0.4 (–)	2.5 (–)	9.8 (–)	7.69 (–)	11.9 (–)	29.39 (–)		
Enhydra fluctuans	0.6 (–)	0.4 (–)	1.5 (–)	5.88 (–)	7.69 (–)	7.14 (–)	22.71 (–)		
Eriocaulon sp.	2.2 (0.8)	0.8 (0.4)	2.75 (2.0)	21.57 (16.66)	15.38 (15.38)	13.09 (14.81)	50.04 (46.85)		
Fimbristylis sp.	— (0.6)	— (0.4)	— (1.5)	— (12.5)	— (15.38)	— (11.11)	— (38.99)		
Hedyotis corymbosa	0.4 (–)	0.2 (–)	2.0 (–)	3.92 (–)	3.85 (–)	9.52 (–)	17.29 (–)		
Hydrocotyle rotundifolia	0.6 (0.4)	0.6 (0.4)	1.0 (1.0)	5.88 (8.3)	11.54 (15.38	4.76 (7.41)	22.18 (31.12)		
Lindernia antipoda	2.2 (1.4)	0.8 (0.4	2.75 (3.5	21.57 (29.16)	15.38 (15.38)	13.09 (25.92)	50.04 (70.46)		
Lippia nodiflora	0.6 (–)	0.4 (–)	1.5 (–)	5.8 (–)	7.69 (–)	7.17 (–)	20.71 (–)		
Panicum sp.	1.2 (0.6)	0.6 (0.4)	2.0 (1.5)	11.76 (12.5)	11.54 (15.38)	9.54 (11.11)	32.82 (38.99)		

Abbreviations: Den., density; Freq., frequency; Abun., abundance; Rel. den., relative density; Rel. Freq., relative frequency; Rel. Abun., relative abundance. After [47].
Average and standard errors cannot be determined through the abovementioned indices.

Table 2.
Phytosociological analysis, Shannon index (H), Simson's index (D) of diversity of the weeds in plots irrigated with pond water and arsenic-contaminated STW waters.

and sodium arsenite [48]. It was suggested that because of the so-called soil/plant barrier effect, elevated arsenic in soil may well reduce crop production substantially before enhanced food chain accumulation occurred [49].

Of the 4 million ha irrigated crop fields, 75% of fields use 100% GW [10]. A significant amount of arsenic withdrawn from underground remains as soil arsenic. Monsoon rain and flood waters are washing away As which is carried to the estuaries at the end and is being accumulating there year after year. This would affect the marine flora and fauna in the near future [47]. A significant amount of As (0.86 (SE 0.057; CV 34.66) mg As/kg) was absorbed by the leaves of *Sonneratia apetala*, a mangrove plant in three coastal islands of Bangladesh, indicating that groundwater As is being accumulated in the biotic and abiotic components along the coast as well [49, 50].

In 1997 the presence of arsenic in all coastal districts, in different level of toxicities—highly contaminated (Bagerhat and Noakhali), moderately contaminated (Sundarbans and Lakshmipur), and low contaminated (Pirojpur, Patuakhali, Bhola, and Feni)—was reported [1] (**Figure 1**). Various studies showed increased arsenic in all types of tube wells as they become older (**Table 1**), indicating that the toxicity levels seen in the mid-1990s might have increased from low to moderate, moderate to high, or even absence of arsenic in the rest of the districts to low or moderate presence of arsenic in over the last 20 years. However, status of arsenic throughout Bangladesh including sites mentioned in [1] should be thoroughly checked.

3.5 Arsenic toxicity management

There are two methods to manage arsenic toxicity in arsenic-loaded DW: first producing biogas and second using as animal feed.

3.5.1 Use in biogas production

The arsenic-loaded DW can be predigested for 4–5 days in summer months covering in polythene. The predigested 60–70 kg DW and similar amount of cow dung can be charged into a doom-shaped biogas digester buried into the soil. The retention time can be 50 days. Everyday 25 kg *S. polyrhiza* and 10 kg cow dung can be mixed and then added into the digester. The gas contains 63–65% methane compared to 60% by cow dung only. Three m^3 volume of biogas can be produced daily, sufficient for cooking twice by a family with eight members. The slurry containing concentrated arsenic can also best be disposed by making bricks binding/trapping the arsenic for over 100 years [28]. This approach requires additional huge structural investment.

3.5.2 Use as animal feed

S. polyrhiza could be used to feed fish, poultry, and cattle [29]. The fresh duckweed is used directly as fish feed keeping at corners within floating fences of a pond. The fish production was 20% higher than the normal feed (long-term effect of toxicity was not studied). The duckweed powder at 4% mixed with the normal feed of broilers per day caused accumulation of 1.28 μg/l arsenic in the blood that reduced to 0.912 ± 0.386 μg/l after 3 months, while in the same period, 4.67 μg/kg stool gradually decreased to 0.61 μg/kg. Feeding arsenic-loaded duckweed to goat resulted in increased accumulation of As (0.567–1.060 μg/kg) in the blood causing death of kids [28]. Milk had also high As (0.86 μg/l). Thus the use of arsenic-loaded DW produced in the bio-mitigation process as poultry and cattle feeds was highly hazardous [29].

3.5.3 Status of As toxicity management

Bioremediation process for getting As-free GW for irrigation that was tested needs additional huge investment (two ponds, DW production cost, cost of labor, etc.). Moreover, an estimated 1.539 g As will remain in the bottom of the pond after treatment. Disposal of the arsenic-loaded DW to nontoxic level through various uses was hazardous (Section 3.5.2). STW and DTW waters were found to become arsenic contaminated in 2–5 years. Withdrawal of the GW contaminates biosphere permanently, i.e., circulating the element in the nature in a matter of weeks or months affecting biotic components (**Figure 3a–d**). Therefore, alternative arsenic-free freshwater sources (about 1000 l/m^2/year equivalent to 2500 l or 2500 kg water/kg arsenic-free rice cultivation in clayey soil) have to be managed for feeding millions of people of Bangladesh on the one hand and saving our biosphere on the other hand. It is possible only by using surface water (Section 3.6.3).

3.6 Current status of arsenic management in Bangladesh

3.6.1 Occurrence and level of arsenic toxicity in Bangladesh

There are 64 districts, out of which 26 were arsenicosis affected in various degrees surveyed between 1993 and 1997 [1] (**Figure 1**): (a) highly affected (>100 patients identified) in 7 districts, (b) moderately affected (>50 patients identified) in 11 districts, (c) less affected (<50 patients identified) in 8 districts, and (d) arsenic contamination present (no patient was identified) in 18 districts. Out of 64 districts, 42 were distributed in four floodplains: (a) Ganges had 26 districts (7 highly affected), (b) Meghna 10 districts (1 highly affected) and (c) Surma-Kushiara 5 districts with arsenic presence but no patients were found, (d) Jamuna 2 districts without any patient, and (e) 2 districts in Madhupur tract (less affected) all due to drinking and consuming arsenic-contaminated GW and food grains/vegetables. Two-thirds of the districts/country was affected with arsenic by the year 1997 [1]. Highly affected Bagerhat and moderately affected major part of Sundarbans south of the Ganges floodplain along the coast are alarming, indicating coastal water pollution. Data on arsenic in irrigation water and paddy soil profiles in Bangladesh [16] and West Bengal [51] indicated a yearly input of 1.0 and 1.1 mg As/kg soil, respectively, in the topsoil (soil density of 1 kg/l). Therefore, distribution of arsenicosis reported in 1997 [1] would be much higher over the last two decades through arsenic toxicity of rice, wheat grains/vegetable, etc. if not by arsenic-contaminated drinking water [17].

Why the Ganges and Meghna floodplains are so much affected with GW arsenic, when Jamuna floodplain with one of the largest and longest river is not? Is it that the GW of the two highly affected floodplains has some link with arsenic mines/industries or there are anthropogenic reasons (dumping the contaminant, deep into the aquifer) in upstream?

3.6.2 Intensity of irrigation

The total area under irrigation in Bangladesh is 4 million ha, and 75% is covered by GW resources: 2.5 m ha via 924,000 STWs (main source of GW As) and 0.6 m ha via 23,000 DTWs [11]. DTW for irrigation is installed at about 100 m depth, and in Jessore alone 74 among 85 DTW used for irrigation had >50 µg As/l arsenic [51]. The rest 25% land is irrigated using surface water of rivers, "Beels," "Haors," etc. In dry season, 3.5 m ha is used for Boro rice, 0.23 m ha for wheat, and 2.7 m ha for other crops. Rajshahi Division has the highest percentage under irrigation which is 39%, followed by Dhaka 27%, Chittagong 13%, Khulna 12%, Sylhet 7%, and Barisal 2% [11]. We must step up the use of surface water from 25% to 100% (Section 3.6.3).

3.6.3 Methods for arsenic's reduction

It has been clearly indicated that As concentrations in rice are increasing over time because of prolonged input of As-contaminated irrigation water, and three options are proposed to free rice grains from toxicity: reduce As-contaminated irrigation water use in rice cultivation, promote cropping patterns, and select/breed rice cultivars that are tolerant to As and have limited uptake of As [11]. As per the first option of limited As-contaminated irrigation water use, there must be alternative sources of huge surface water, e. g., initially forming reservoirs and constructing rubber dams in rivers, and execution of long-term "Delta Plan 2100". Regarding the second option, cropping pattern throughout the country has been established over the decades of testing, while selection/breeding of rice varieties tolerant to As and limited uptake are questionable. It is known that As and P elements are in the same position in the periodic table (chemically similar), and thus the rice cultivar that will not absorb As will not absorb P as well. In the present study, arsenic bio-mitigation of irrigation water tested was not effective, and disposal of wastes was hazardous. As arsenic-contaminated GW produces toxic rice grains and accumulates arsenic in the soil year after year, avoiding the use of GW is the only solution for protecting rice grains from arsenic toxicity, including other organisms. Man might alter the quantities of arsenic in any component of an ecosystem in a localized area but cannot change or stop the natural biological processes that occur [52]. Therefore, an alternative immediate attention is needed to provide enormous volume of As-free irrigation water, and that is through the use of surface waters using river network of Bangladesh (**Figure 10**). Bangladesh is a country of rivers having almost one river in each village. The rivers have to be dredged to get continuous flow of waters from the upstreams, converting the rivers as reservoirs simultaneous with the construction of rubber dams. Several rubber dams have so far been installed in Bangladesh, one of which is at Sonargaon (**Figure 11**). Bangladesh government has the plan to establish **"Ganges Barrage"** to supply freshwater to Ganges floodplain and surroundings including Sundarbans [53]. Huge deposit of sediments in the Ganges River bed has been identified as a major problem (what to do with the sediments) for using the Ganges as a reservoir [personal communication, Rawshan Ali Khan, Project Director of Ganges Barrage Project, Dhaka, Bangladesh].

The National Economic Council (NEC) of Bangladesh has recently approved **"Delta Plan 2100"** the key objective of which is to provide food and water security and fight natural disasters [54]. The theme is "let the rivers flow, let the rivers live." In the first phase, the government will be implementing about 80 projects at an estimated cost of US $ 37 billion by 2030. Out of six goals, Delta goal 3 is "Ensuring sustainable and integrated river systems and estuaries." In a seminar, it has been mentioned that "rivers would be channelized and sediments would be removed." Details are not available regarding what is meant by the "removal." It immediately indicates dumping sediments on to the river banks!

The landmass of Bangladesh has been formed throughout the Pleistocene and up to the present by sediments washed down from the Himalaya Mountains through the Ganges, Jamuna (Brahmaputra), and Meghna Rivers and their numerous tributaries and distributaries [55]. In terms of relative age of the landmass, the region may be divided into four parts: hilly lands of the Tertiary (and older) in the southeast Chittagong and CHT districts, terrace lands of

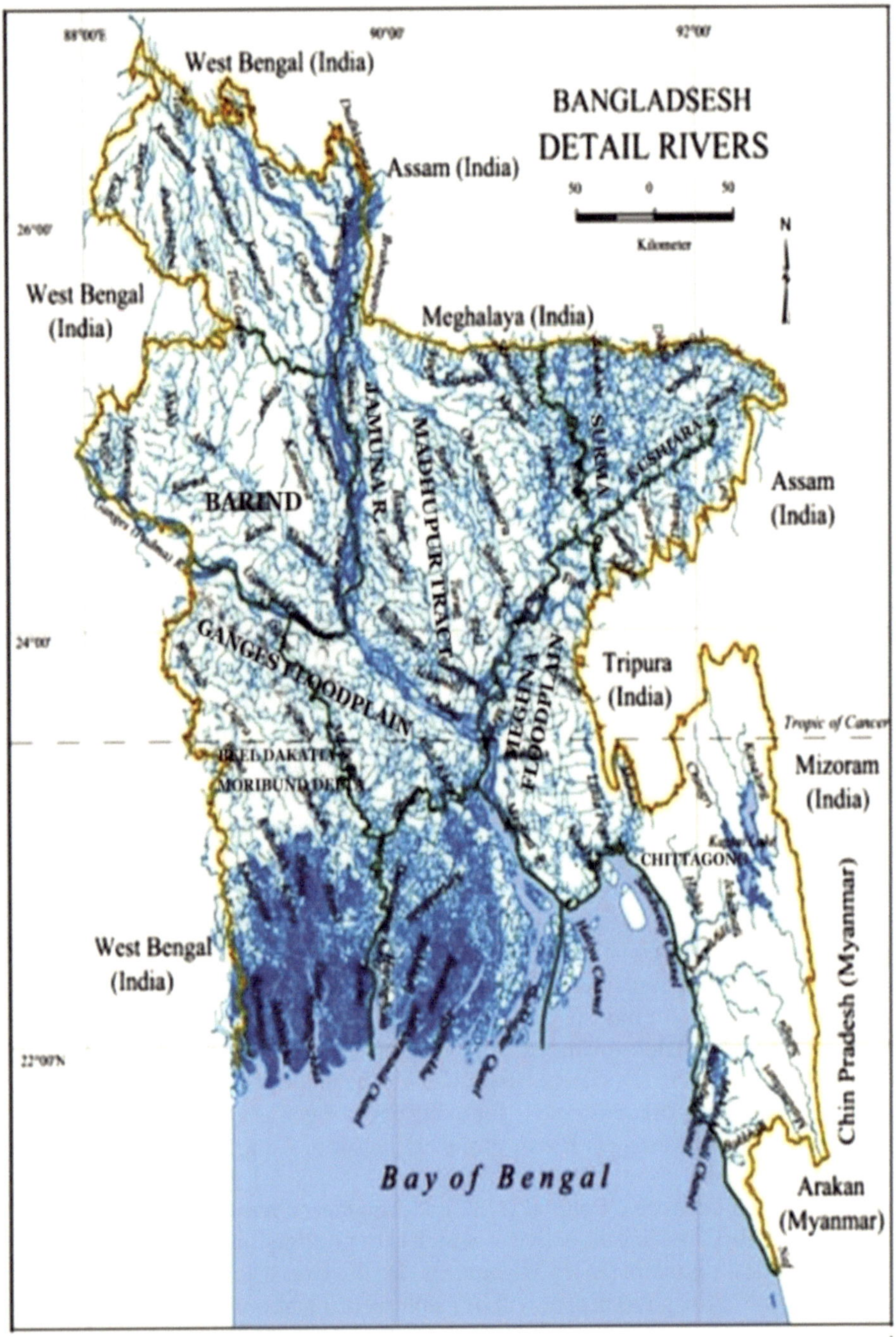

Figure 10.
River network of Bangladesh, where three mighty rivers Ganges, Jamuna (Brahmaputra), and Meghna altogether ended into the bay of Bengal.

Pleistocene in the Barind and Madhupur Jungle, tipper surface of the early Recent in the median eastern part, and the extensive floodplains of the Recent in the rest of the country (Ganges floodplain and southern part of Meghna floodplain) (**Figures 1** and **10**). Geologists believe that in the same way, the present Sundarbans has been established 7000 years ago [56]. From geophysical

Figure 11.
Rubber dam installed in the old Brahmaputra River at Sonargaon, Bangladesh, in 2003. (a) Showing low water volume in the river bed at noon time, on the north of bridge, and(b) rubber dam inflated (on the left of the bridge) to hold water in the river during afternoon.

changes on the formation of Bangladesh, it is obvious that sediment-loaded river water flow to the Bay of Bengal resulted from extensive floodplains including Sundarbans [56], and any disruption would cause land degradation, which is already done by Farakka Barrage since 1975 [53]. To reclaim environment of all the floodplains, dredging of Ganges, Jamuna, and Meghna River beds along with tributaries and distributaries making them flow round the year is primarily needed and letting the dredged sediments flow along with floodwater during June to October, up to the Bay of Bengal", the way Bengal Delta was formed [53, 56]. The surface water is ideal in the sense that it contains all the nutrient elements in relatively the same proportion (Redfield ratio) for irrigating rice and other crops and protecting crops and biota from arsenic toxicity. Increased surface water would reduce GW salinity in the southwestern Bangladesh and increase productivity in the hinterlands [53]. To get rid of arsenics, protecting surface water is the only environmentally benign option saying "no to groundwater for irrigation, let the arsenic stay in the underground." The water reservoirs may integrate aquaculture of DW and *A. pinnata* var. pinnata for fish and poultry feeds [57].

4. Summary

The arsenics causing arsenicosis in Bangladesh water were first detected in 1993. Presently, many districts of Bangladesh use As-contaminated GW for rice irrigation and an integrated approach within the framework of land degradation has been suggested [11]. Considering the necessity of huge volume of arsenic-free irrigation water, physiography of Bangladesh and three mighty rivers flowing through the Bay of Bengal forming extensive network of rivers within the country, immediate attention must be given to construct rubber dams, convert rivers into freshwater reservoirs to overcome the arsenic toxicity, and protect rice and other grains. The integrated approach described in this manuscript would be environment friendly increasing total crop productivity including aquaculture of duckweeds and A. *pinnata* var. *pinnata* for producing fish and poultry feeds.

Permanent solution could be achieved executing "Delta Plan 2100" saying "No to groundwater use for irrigation, let the Arsenic stay in the underground".

Acknowledgements

The author is grateful to the Ministry of Science and Information & Communication Technology, Govt. of Bangladesh, for supporting the research project entitled "Mitigation of arsenic poisoning of groundwater by employing a locally innovated bioremediation technique," no. MOSICT/Sec. 9/SAP 3/2002-2003. The author is thankful to Dr. Syed Mohammad Hossain, Institute of Nuclear Science and Technology, Atomic Energy Research Establishment, Saver, Dhaka, for his help in determining arsenic in rice grains by neutron activation analysis.

Author details

Abdul Aziz
Department of Botany, University of Dhaka, Dhaka, Bangladesh

*Address all correspondence to: dr.aziz.botany@gmail.com

References

[1] Khan AW, Ahmad SA. Arsenic in Drinking Water, Health Effects and Management: A Training Manual. Dhaka: Department of Occupational and Environmental Health (DOEH), National Institute of Preventive and Social Medicine (NIPSOM); 1997. p. 47

[2] Bridge TH, Hosain MT. Groundwater arsenic poisoning and a solution to the arsenic disaster in Bangladesh. In: Intl. Conf. Bangladesh Environ; 14-15 January 2000; Dhaka: BUET. 2000. pp. 95-105

[3] BGS (British Geological Survey). Groundwater Studies for Arsenic Contamination in Bangladesh. Review of Existing Data. Vol. S1. United Kingdom: Mott MacDonald Ltd; 1999. pp. 1-33

[4] Walsh LM, Keeney DR. Behaviour and phytotoxicity of inorganic arsenicals in soil. In: Woolson EA, editor. Arsenical Pesticides. Washington, DC: ACS; 1975. pp. 35-52

[5] Dey RK. Arsenic health problem. In: Ahmed MF, Ahmed CM, editors. Arsenic Mitigation in Bangladesh. Dhaka: Local Govt. Division; 2002. pp. 59-76

[6] Anderson AC, Abdelghani AA, McDonell D. Screening of four vascular plants for uptake of monosodium methane arsenate (MSMA). Science of the Total Environment. 1980;**16**:95-98

[7] Wauchope RD. Uptake, translocation and phytotoxicity of arsenic in plants. In: Lederer WH, Fensterheim RJ, editors. Arsenic: Industrial, Biomedical and Environmental Perspectives. NY: Van Nostrand Reinhold; 1983. pp. 348-377

[8] Jain SK, Vasudevan P, Jha NK. Removal of some heavy metals from polluted water by aquatic plants: Studies on duckweed and water velvet. Biological Wastes. 1989;**28**:115-126

[9] Aziz A. Removal of arsenic from water by *Azolla* and *Spirodella*. I. Journal of the Asiatic Society of Bangladesh, Science. 2000;**26**:271-276

[10] Mamtaz R, Bache DH. Low-cost technique of arsenic removal from water and its removal mechanism. In: Ahmed MF, Ali MA, Adeel Z, editors. Technologies for Arsenic Removal from Drinking Water. Dhaka: BUET and United Nations University; 2001. pp. 43-58

[11] Heikens A. Arsenic Contamination of Irrigation Water, Soil and Crops in Bangladesh: Risk Implications for Sustainable Agriculture and Food Safety in Asia. UN: Regional Office for Asia and the Pacific, FAO; 2006. p. 38

[12] Zhu Y-G, Williams PN, Meharg AA. Exposure of inorganic arsenic from rice: A global health issue? Environmental Pollution. 2008;**30**:1-3

[13] Das HK, Mitr AK, Sengupta PK, Hossain A, Islam F, Rabbani HH. Arsenic concentration in rice, vegetables and fish in Bangladesh: A preliminary study. Environment International. 2004;**30**:383-387

[14] Juhasz AL, Smith E, Weber J, Rees M, Rife A, Kuchel T, et al. *In vivo* assessment of arsenic bioavailability in rice and its significance for human health risk assessment. Environmental Health Perspectives. 2006;**114**:1826-1831

[15] Xu XY, Mcgrath SP, Meharg AA, Zhao FI. Growing rice aerobically markedly decreased accumulation.

Environmental Science & Technology. 2008;**42**:5574-5579

[16] Meharg AA, Rahman M. Arsenic contamination of Bangladesh paddy field soils: Implications for rice contribution to arsenic consumption. Environmental Science & Technology. 2003;**37**:229-234

[17] Meacher DM, Menzel DB, Dillencourt MD, Bic LF, Schoof RA, Yost LJ, et al. Estimation of multimedia inorganic arsenic intake in the US population. Human and Ecological Risk Assessment. 2002;**8**:1697-1721

[18] Jacobs LW, Keeney DR, Walsh LM. Arsenic residue toxicity to vegetable crops on Plainfield sand. Agronomy Journal. 1970;**62**:588-591

[19] Islam MA, Ullah SM. Effects of arsenic on some growth parameters of rice plants. Dhaka University Journal of Biological Sciences. 2003;**12**:71-77

[20] Islam MA, Ullah SM, Siddique AB. Effects of arsenic on growth and nutrient uptake in cowpea plants (*Vigna sinensis* Prain). Dhaka University Journal of Biological Sciences. 2004;**13**:103-106

[21] Jahan I, Hoque S, Ullah SM, Kibria MG. Effects of arsenic on growth and nutrient uptake in rice plants (*Oryza sativa* L.). Dhaka University Journal of Biological Sciences. 2003;**12**:15-22

[22] Jahan I, Hoque S, Ullah SM, Kibria MG. Effects of arsenic on some growth parameters of rice plant. Dhaka University Journal of Biological Sciences. 2003;**12**:71-77

[23] Zingaro RA, Bottino NR. Biochemistry of arsenic: Recent developments. In: Lederer WH, Fensterheim RJ, editors. Arsenic: Industrial, Biomedical and Environmental Perspectives. NY: Van Nostrand Reinhold; 1983. pp. 327-347

[24] Bhumbla DK, Keefer RF. Arsenic mobilization and bioavailability in soil. In: Nriagu JO, editor. Arsenic in the Environment, Part 2. London: Wiley; 1994. pp. 51-81

[25] Nordstrom DK. An overview of arsenic mass-poisoning in Bangladesh and West Bengal, India. In: Young CA, editor. Minor Metals 2000. Littleton, CO: SME; 2000. pp. 21-30

[26] WHO. Trace Elements in Human Nutrition and Health. FAO. 1996. p. 361

[27] Johansorton R, Heisen H. Safe water technology for arsenic removal. In: Ahmed MF, Ali MA, Adeel Z, editors. Technologies for Arsenic Removal from Drinking Water. International Workshop; May 2001. Dhaka: Bangladesh Univ. Engr. Tech and United Nations University; 2001. pp. 1-23

[28] Robins RG, Nishimura T, Singh P. Removal of arsenic from drinking water by precipitation, adsorption or cementation. In: Ahmed MF, Ali MA, Adeel Z, editors. Technologies for Arsenic Removal from Drinking Water. Dhaka: BUET and United Nations University; 2001. pp. 31-42

[29] Aziz A, Ullah SM, Chowdhury A, Alam M. Mitigation of arsenic poisoning of ground water by employing locally innovated bio-remediation technique. In: Final Report (Mimeo), Ministry Sci. Info. And Comm. Tech., GOB. 2004. p. 42

[30] Aziz A, Whitton BA. Morphogenesis of blue-green algae. I. Filament development in *Gloeotrichia*. Bangladesh Journal of Botany. 1987;**16**:69-81

[31] Chu SP. The influence of the mineral composition of the medium on the growth of planktonic algae. I. Methods and culture media. Journal of Ecology. 1942;**30**:284-325

[32] Watanabe I, Roger PA, Ladha JK, Van Hove C. Biofertilizer Germplasm Collection at IRRI. Manila: IRRI News Letter; 1992

[33] Alaerts GJ, Rahman MM, Kelderman P. Performance analysis of a full-scale duckweed-covered sewage lagoon. Water Research. 1996;**30**:843-852

[34] Aziz A, Kochi MN. Growth and morphology of *Spidodela polyrhiza* and *S. punctata* (Lemnaceae) as affected by some environmental factors. Bangladesh Journal of Botany. 1999;**28**:133-138

[35] Aziz A, Sharmin S. Growth and nitrogenase activity of *Azolla pinnata* var. *pinnata* as affected by some environmental factors. Bangladesh Journal of Botany. 2000;**29**:126-131

[36] Hardy RWF, Burns RC, Holsten RD. Application of the acetylene-ethylene assay for measuring nitrogen fixation. Soil Biology and Biochemistry. 1973;**5**:47-81

[37] Eisenreich SJ, Bennerman RT, Armstrong DE. A simplified phosphorus analysis technique. Environmental Letters. 1975;**9**:43-53

[38] USDA. Foreign agricultural service global agriculture information network report CH6064. In: FAIRS Product Specific Maximum Levels of Contaminations in Foods; People's Republic of China. 1951. p. 147

[39] USSL. Diagnosis and improvement of saline and alkali soils. In: Agric. Handbook No. 60, USDA. 1954. p. 147

[40] Piper CS. Soil and Plant Analysis. Adelaide: The Uni. Of Adelaide; 1950. p. 368

[41] Black CA. Methods of Soil Analysis, Part II. Wisconsin Am. Soc. Agron. Inc. Publ; 1965. p. 1188

[42] Day PR. Particle fractionation and particle size analysis. In: Black CA, Evans DD, Ensminger LE, White JL, Clark FE, editors. Methods of Soil Analysis. Part 1. Madison, WI; 1965. pp. 545-567

[43] Aziz A, Ullah SM, Ullah MR. Arsenic in rice grains at Sonargaon, Bangladesh. Bangladesh Journal of Botany. 2015;**42**:85-89

[44] Norra S, Berner ZA, Agarwala P, Wagner F, Chandrasekharam D, Stuben D. Impact of irrigation with As rich groundwater on soil and crops: A geochemical case study in West Bengal Delta plain, India. Applied Geochemistry. 2005;**20**:1890-1906

[45] Duxbury JM, Zavala YJ. What Are Safe Levels of Arsenic in Food and Soils? Cornell University; 2010. p. 5

[46] Meharg AA, Sun G, Williums PN, Adomako E, Deacon C, Zhu Y-G, et al. Inorganic arsenic levels in baby rice are of concern. Environmental Pollution. 2008;**152**:746-749

[47] Aziz A, Ahmed A. Effects of arsenic on diversity of rice field weeds. Dhaka University Journal of Biological Sciences. 2010;**19**:95-98

[48] Chandrakar V, Keshavkant S. Growth and metabolic responses of *Glycine max* L. to arsenate and arsenite: A comparative assessment. Bangladesh Journal of Botany. 2018;**47**:105-113

[49] Cheney RL. Potential effects of sludge-borne heavy metal and toxic organics on soil, plants, and animals and related regulatory guidelines. In: Proc. Pan American Health Organization Workshop on the International Transportation, Utilization or Disposal of Sewage Sludge, Including RecommendationsFinal Report PNSP/85-01; 12-15 December.

Washington, DC: Pan American Health Organization; 1984. pp. 1-56

[50] Ahmed A, Ohlson M, Hoque S, Moula MG. Chemical composition of leaves of a mangrove tree (*Sonneratia apetala* Buch.-Ham.) and their correlation with some soil variables. Bangladesh Journal of Botany. 2010;**39**:61-69

[51] JICA/AAN. Arsenic Contamination of Irrigation Tube Wells in Sharsha Upazila, Jessore. Dhaka: Asia Arsenic Network; 2004. p. 13

[52] Woolson EA. Man's perturbation of the arsenic cycle. In: Lederer WH, Fensterheim RJ, editors. Arsenic: Industrial, Biomedical, Environmental Perspectives. NY: Van Norstrand Reinhold; 1983. pp. 393-408

[53] Aziz A. Bangladesh Sundarbans: Present status of the environment and biota. Diversity. 2015;**7**:242-269

[54] The Daily Star. Ensuring Free Flow of Rivers Key Objective of Delta Plan; 6 September 2018; XXVIII No. 226; Dhaka. Email: editor@thedailystar.net

[55] Morgan JP, McIntire WG. Quaternary geology of Bengal Basin, East Pakistan and India. Bulletin Geological Society of America. 1959;**70**:319-342

[56] Choudhuri AB, Choudhury A. Mangroves of Sundarbans. Vol. 1. Bankok, India: IUCN; 1994. p. 247

[57] Aziz A. Arsenic problems: Integrated approach for solution by floating plants. In: Ahmed MF, Tanveer SA, Badruzzaman ABM, editors. Bangladesh Environment. Vol. 1. Dhaka: Bangladesh Poribesh Andolon; 2002. pp. 186-199

Chapter 10

Smart Nutrition Management of Rice Crop under Climate Change Environment

Rai Mukkram Ali Tahir, Noor-us-Sabah, Muhammad Afzal, Ghulam Sarwar and Ijaz Rasool Noorka

Abstract

Soil fertility and plant nutrition remained main pillars of agricultural sciences in twentieth century. However, due to recent interest in achievement of sustainability and restricted natural resources, importance of soil fertility and plant nutrition is expected to be increased many folds in twenty-first century. Therefore, increasing rice crop yield under such scenario will require judicious and efficient use of mineral sources of nutrient with combination of natural resources, recycling of bioavailable nutrients, and genetic modification of crops for efficient nutrient utilization. There is an increasing pressure on agricultural land to produce sufficient amount of food needed to feed the growing global population. The pressure is associated with changing weather patterns related to fluctuations in rainfall and temperature, supply of fertilizers inflating price associated with energy demand, which is very closely linked with weather patterns and reducing soil fertility. Increasing rice yield under these constraints will require a rational use of chemical fertilizers with increase the use of natural resources of nutrition, recycling of plant available nutrients, and an exploitation of the genetic potential of crop species to make efficient use of nutrients a key feature to establish smart plant nutrition management in the recent global climate change scenario.

Keywords: rice, mineral nutrition, aerobic, flooded soil, climate change

1. Introduction

Rice is the world dominant food crop, that is, widely adapted to a variety of climatic zones (temperate, tropical, sub-tropical and semiarid) in all continents [1]. Since differences exist in growth conditions and yields of upland and low land rice resultantly their nutritional requirements are also different [2]. In case of low land rice due to conducive environmental and growth conditions, crop yield positively respond to application of fertilizers. On the other hand, for upland rice, inadequacy of water especially at the stage of flowering resulted in significant yield reductions and crop yield response to application of fertilizers in not evident is the world dominant food crop widely distributed throughout the tropical, subtropical, semiarid and temperate zones of all continents [3]. This crop is dominant due to two major factors as more breeding work done for it and secondly grown mostly on better irrigated land. Based on land and water management, rice culture is

IntechOpen

categorized as upland and low land crop. Upland rice refers to rice grown on both flat and sloping field that are prepared and seeded under dryland conditions and depend on rainfall for moisture. This is also termed as dryland, rainfed and aerobic rice. Low land rice is grown on flat land with controlled irrigation, it is also known as flooded, irrigated and water logged soil. About 76% of the world rice production comes from irrigated area [3, 4].

Nutrient requirements for upland and lowland rice are different due to differences in yield levels and growing conditions. In low land rice, environmental conditions are most stable and favorable for plant growth and high fertilizer application can ensure high yields. But in case of upland rice, inadequate water particularly around flowering mostly reduced yield significantly and at the end there is little or no differences in yield between well fertilized and unfertilized crops [3, 5].

Soil fertility is the major constraint in upland rice production, as rice grown in naturally drained soil without surface water accumulation. Soil acidity, low cation exchange capacity, and high P fixation capacity are the major soil chemical properties affecting upland rice production [3]. In the era of climate change, the carbon dioxide content increased in the atmosphere this impact lead to less nutritious rice a serious issue for human kind due to huge consumption of rice as food, so it's a dire need of time to recommend a precise amount of mineral nutrient in growth environment for healthy growth of rice in this climate change situation. The basic premise of this chapter was to highlight the smart mineral nutrition management approach for the growth of rice under aerobic and flooded soil condition.

2. Nutrient stress

Nutrient stresses refer to deficiencies of essential plant nutrients as well as toxicities. Nutrient deficiencies are more common than toxicities in many arable lands around the world as mentioned in **Table 1** [4].

In the 1980s and 1990s evidence accumulated that nutrient depletion is problem in many tropical soils. If nutrient stress is not alleviated, crop yields are decreased and soils cannot support adequate plant growth. Under this situation, soil degradation starts. If the land continues to be used for crop production, crop yields become so low that farmers have to abandon the degraded areas. Approximately 1/4th of the earth's soils are considered to produce some kind of mineral stress in crops [3, 4, 6].

2.1 Nutrient stresses alleviation system

For the assessment of efficient working of nutrient management practices on sustainable basis, soil testing need to be done frequently. Soil testing is one of the keys to success of cost-effective, environmentally benign and effective sustainable farming program [4]. Soils on which rice grows varied in texture, climate, pH, salt content, organic matter content and nutrient availability [7]. Fertilizer recommendation for a given crop should be based on series of research trials, the results should then fit on response functions and then profitability should be calculated by using economic variables and equations in order to calculate optimum doses of fertilizers. Following outcomes are achieved by adopting adequate soil fertility in terms of prevention of land degradation.

i. Adequate fertility gives the crop a better vigor at early stages and good canopy cover which protect the soil from erosion.

Element	Deficiency (D)/toxicity (T)	Critical level	Plant part	Growth stage
N	D	2.5%	Leaf blade	Tillering
P	D	0.1%	Straw	Maturity
	T	1.0%	Straw	Maturity
K	D	1.0%	Straw	Maturity
	D	1.0%	Leaf blade	Tillering
Ca	D	0.15%	Straw	Maturity
Mg	D	0.10%	Straw	Maturity
S	D	0.10%	Straw	Maturity
Fe	D	70 ppm	Leaf blade	Tillering
	T	300 ppm	Leaf blade	Tillering
Zn	D	20 ppm	Shoot	Tillering
	T	1500 ppm	Straw	Maturity
Mn	D	20 ppm	Shoot	Tillering
	T	2500 ppm	Shoot	Tillering
B	D	3.4 ppm	Straw	Maturity
	T	100 ppm	Straw	Maturity
Cu	D	6 ppm	Straw	Maturity
	T	30 ppm	Straw	Maturity
AI	T	30 ppm	Shoot	Tillering

Table 1.
Deficiency and toxicity of essential plant nutrients along with critical level content at different growth stages of rice plant.

ii. After crop harvesting provision of more crop remains on soil surface provide the soil protection against wind and water erosion and buildup of soil organic matter status thereby, increasing soil potential for crop raising on long term basis.

iii. An improvement efficient utilization of nutrients by crops results due to adequate fertility, which gives both economic and environmental benefits. Balanced NPK fertilization results in enhancement of nitrogen utilization efficiency. Thereby, mitigating nitrogen losses due to leaching and safe guard the ground waters from nitrate pollution.

iv. Adequate soil fertility results in water conservation by increasing crop water use efficiency.

v. Adequate soil fertility maximize the outcomes from crop due to positive interaction with other production inputs like tillage practices, pest control management, selection of crop variety etc.

Under the scenario of global climate change, placement of fertilizer is also integral component of efficient crop management in addition to adequate rate of fertilizer. Placement of fertilizers has role in nutrient utilization and ultimately on crop yield. Placement of nutrients (N, P and K) in the form of band is preferred over broad casting due to many reasons, for example,

- In case of band placement fertilizer is placed in the vicinity of roots thus making the nutrients readily available to plants.
- As compared to broadcasting less quantity of fertilizer is used in band placement.
- Due to placement of nutrients at right place there is no weed emergence therefore lowering the cost of weedicide or weeds eradication.
- Crop nutrient use efficiency is improved in case of band placement.
- Nutrient losses resulting from erosion, immobilization and leaching of N is significantly reduced by adopting band placement technique.

Implementation of this system to any cropping system results in optimization of nutrient management and reduction of soil degradation. In addition improvement in soil nutritional deficiencies and reduction in land degradation also be achieved by the exploitation of genetic variability of plants in terms of nutrient absorption and utilization. Efforts are being made around the globe to make use of genetic resources in order to genetically modify the crop cultivars to produce such cultivars that are more efficient and productive under stressed environments [3–6, 8, 9].

3. Rational nutrient management approaches for rice growth under climate change environment

3.1 Integrated plant nutrient management approach

Integrated plant nutrient management system (IPNS) is a holistic approach to integrate the one of all natural man made sources of plant nutrient to maintain and sustain soil fertility to enhance rice crop productivity in an efficient, environmentally safe, ecologically compatible, socially acceptable and economically viable way. It use both organic and in organic plant nutrients to attain higher crop productivity and prevent soil degradation.

IPNS system keep a balance between nutrients removed by the crop and nutrient added to soil. The smart nutrient management program taking into account the availability of nutrients in all types of soil, crop requirement and other factors, such as, removal of nutrients from the soil by the crop, economics of fertilizer profitability, farmers ability to invest, soil moisture regime, physical and microbiological condition of the soil, available soil nutrient status, nutrient recycling and cropping sequence, limiting loss to the environment [10, 11] (**Figure 1**).

3.2 Four R nutrient management approach in rice crop production

Soil is marvelous, complex substances. Thousands of soil types exist in the world having arisen from different parent material under diverse ecological conditions. Some are fertile, tillable and wonderfully suited for agriculture other may need a great deal of husbandry to become useful. Sustainable agriculture or regenerative farming all aim to produce food and fiber on a sustainable basis and to repair the damage caused by destructive procedure [3, 4] (**Figure 2**).

4R nutrient stewardship is based on the principle that fertilizer should be applied meeting the following requirements:

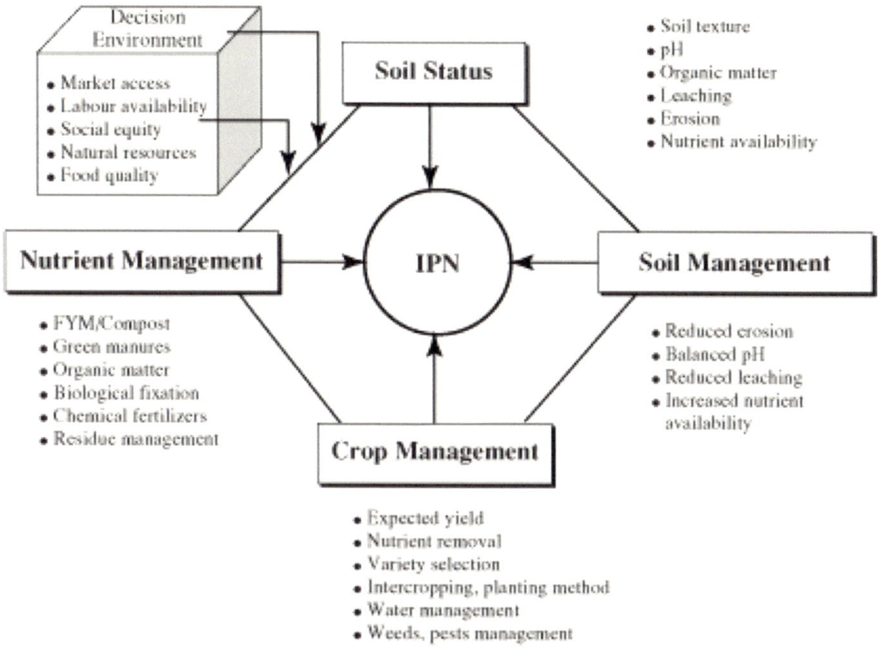

Figure 1.
Integrated plant nutrient management system in industrial agriculture.

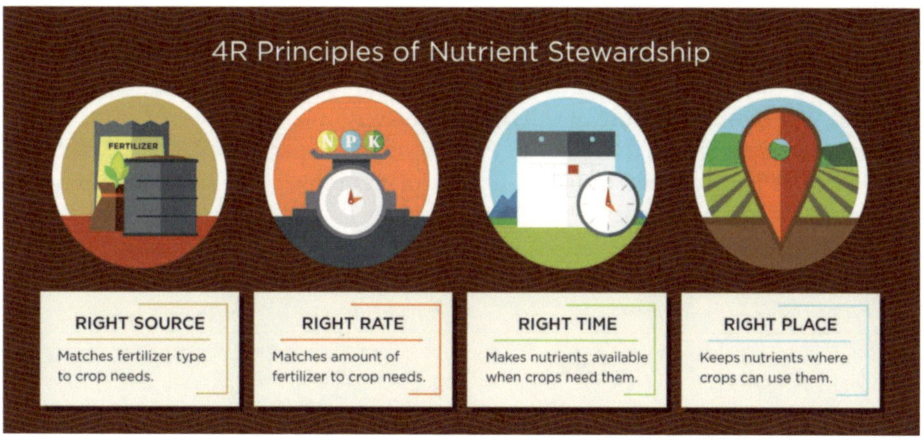

Figure 2.
Four-R nutrient stewardship model for best management practices (BMP).

i. **R**ight type or source

ii. **R**ight rate or dose

iii. **R**ight place and

iv. **R**ight time

Adaptation of this model helps the growers to achieve sustainability. Although the recommendations are specific with respect to site but scientific principles are universal.

4. Smart mineral nutrition management of essential plant nutrients in the scenario of global climate change

4.1 Smart nitrogen fertilizer management

Management of no other fertilizer nutrient presents a great challenge to the rice. The N fertilizer rate required to achieve optimum yield in rice can be influenced significantly by the preceding crop. The nitrogen fertilizer rate required to produce the best grain and milling yield of rice is dependent on rice genotypes, stand density, previous crops, straw management, soil texture, permeability, N fertilizer methods, water management, soil reaction, tillage and N fertilizer source. In rice plant nitrogen fertilizer loss mainly by nitrification and denitrification, and diffusion (**Figure 3**) [8, 9].

Rice grain yield mainly affected by the number of tillers, which, in turn, is influenced by the N fertilizer rates. The application of N for rice highly depends on soil separates content. In sandy soil with low CEC fertilizer is subjected to considerable leaching losses, higher N rate or multiple N applications may be considered to overcome losses. The clay soils generally need 40–60 kg N/ha more N fertilizer than those rice grown on silt loams to achieve similar grain yield. The use of climate smart nitrogen fertilizer like: neem coated, sulfur coated, polyamine coated that may improve yield of rice under tough soil and climate condition [12].

4.2 Smart phosphorous nutrition management

Adequate P nutrition of rice is essential because it is needed for energy storage and transfer within plant body. In rice P ensured early maturity, straw strength, and crop quality and disease resistance. Phosphorus exists in soil in two basic pools, organic and inorganic. The organic P (Po) is the part of soil organic matter and soil biomass. The dynamic nature of soil organic matter mineralization and immobilization processes dictate that some of Po contributes to plant available P. Actually inorganic P (Pi) regulates P nutrition for rice plant uptake [13]. Pi in soils has been characterized by five forms: (1) calcium phosphate (Ca-P), (2) iron phosphate (Fe-P), (3) aluminum phosphate (Al-P), (4) occluded P (O-P), (5) soluble orthophosphate (Sl-P) last once more readily utilized by plant body within a wide range of

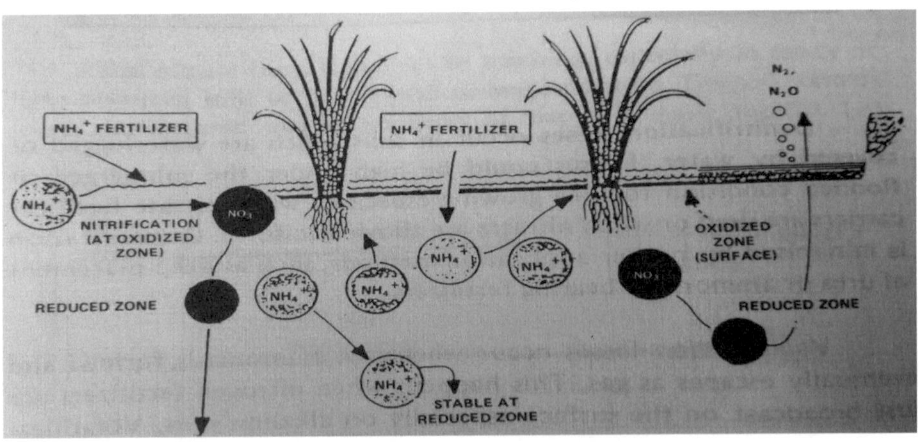

Figure 3.
Nitrogen chemicals forms, transformations, and behavior in the flooded soil environment in which rice is grown. Nitrogen sources are in blocks, nitrogen chemical form are in circles and the mechanisms responsible for the various nitrogen transformations or behavior are located on the arrowed lines [8, 9, 13].

soil reaction. The availability of other forms in different soil system depends upon sorption, desorption and diffusion by maintain equilibrium between labile and solution P levels. P deficiency mostly shown as bronzed leaf very similar symptoms appeared when rice grown on Zinc deficiency soil. In general, P fertilizer rates of 30, 20 and 10 kg P/ha are recommended for rice when the soil test are very low, low and medium in P respectively. In industrial agriculture, the foliar P application may also be recommended at the stage of grain filling [11].

4.3 Smart potassium fertilizer management

Potassium is taken up by rice as K^+ ion. Potassium exists in the soil as four basic forms (1) solution, (2) exchangeable, (3) nonexchangeable, (4) mineral; these four forms of K are all in a state of dynamic equilibrium. Potassium deficiency has not been a common problem in rice but P deficiency enhances the occurrences of disease like, kernel smut, stem, and sheath rot'. In the field condition K status change easily from K-deficient to K-sufficient by interactions of K pools. In general recommendation the foliar application of potassium nitrate produced better results against disease in rice plant [10, 11, 14] (**Figure 4**).

4.4 Smart micronutrients management in rice growth

The metal micronutrients reported to affect the growth of rice under climate change environments are zinc, iron and manganese. Several mechanism in which aerobic and flooded environment influence the availability of trace element by (1) increased solubility of compounds via the dilution effect of excess water, (2) pH changes associated with oxidation-reduction reactions, which can cause nutrients to be transformed to soluble and insoluble forms, (3) increased availability due to mobility of nutrients in the saturated soil. Nutrient plant uptake also affected by temperature change [4]. The use of chelated for micronutrients proved successful to improve yield of rice under the condition of global climate condition. Iron and Mn in the soil conceptually exist in four basic forms: solution Fe and Mn, adsorbed Fe and exchangeable Mn, organic complexed Fe and Mn, and Fe and Mn in primary and secondary minerals. All of these forms of Fe and Mn are in equilibrium with solution Fe and Mn, and the organic complexed forms facilitate their transport in the soil solution and uptake by rice. Unlike Zn, these two metal micronutrients can be reduced in flooded soil and become much more soluble and plant available [5, 15, 16] (**Figure 5**).

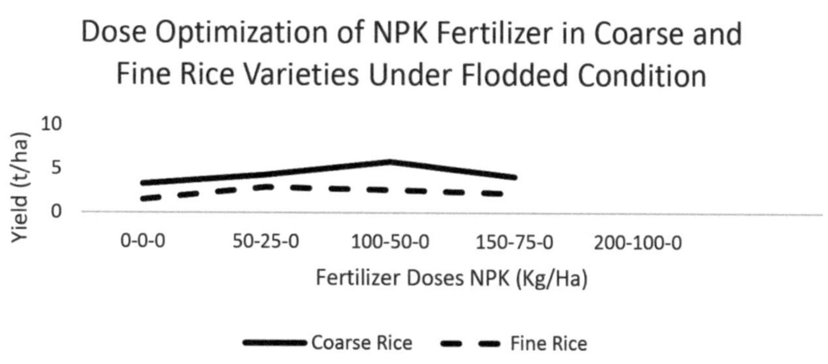

Figure 4.
Dose optimization of NPK fertilizer among two fine and coarse contrasting rice varieties for better yield management.

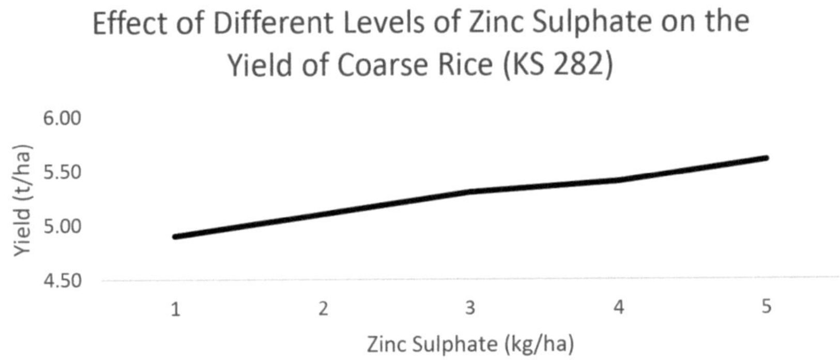

Figure 5.
Influence of commercial grade zinc sulfate on the yield of rice crop.

4.5 Beneficial plant nutrient management and rice growth

Silicon is the second most abundant mineral in the earth's crust but is not considered an essential element for many plant species. Silica is considered a beneficial element for rice growth, because it has not been shown that rice fails to complete its life cycle in the absence of Si. Plant species are categorized as either Si-accumulators or non-accumulators. Rice is considered as Si-accumulator specie. Application of Si amendments have been shown to be beneficial to rice growth, yield, and pest reaction in some areas of the United States. Research in Pakistan has demonstrated that Si containing soil amendments have produced significant yield increase on mineral soils. 2 kg silicon/hectare is required for better yield of rice [10, 11].

5. Conclusion

The change in greenhouse gasses concentration leading to global climate change is evident at present. In this global climate change we can now address plant ability to efficiently use the nutrients available to it whether from chemical fertilizers, manures or what is naturally available in the soil and water by inventing a simple hypothetical perfect food plant which contains high nutritional value of food that we eat.

The aim of this contribution is firstly to print out of some if presently available plant nutrition ways to rice crop and secondly to demonstrate the impact of global climate change on the rice crop growth. We should that efficient use of fertilizer material have to potential combat agent the climate change situation. This enables us to construct a new approach of plant nutrition new concept and methods are desperately needed to achieve the goal of sustainable agriculture growth of rice under alarming global climate change situation. Low productivity in rice crop is mainly caused by native low soil fertility and water stress. The low soil fertility is associated with low organic matter and clay content in soil, which also produce low recovery efficiency of chemical fertilizer.

The field of investigation revealed a linear relationship between paddy yield and fertilizer application rate between N and Zn. Frequent split fertilizer application of slow release fertilizer improved fertilizer recovery efficiency that a key to feature in rice productivity in this climate change scenario.

The integrated use of natural and chemical fertilizer improved the fertilizer holding capacity that critical factor to enhance rice productivity under all type of rice culture. Application of poultry litter, either fresh or composted, has been shown to be the most effective means of quickly restoring rice productivity to soils that have been altered by precision grading. Graded soils have lower organic matter and thus lower amounts of potentially mineralization of native soils N than do typical undisturbed soils. Consequently, N fertilizer rates for rice should be adjusted only when extremely high rates of litter are applied immediately prior to planting.

Global climate change is a major concern in the twenty-first century that may lead to depletion of the soil organic matter that will decrease the efficient uptake of mineral nutrient in rice crop. The smart nutrient management program taking into account the availability of nutrients in all types of soil, crop requirement and other factors, such as, removal of nutrients from the soil by the crop, economics of fertilizer profitability, farmers ability to invest, soil moisture regime, physical and microbiological condition of the soil, available soil nutrient status, nutrient recycling and cropping sequence, limiting loss to the environment. Achieving balance between the nutrient requirements of plant and the nutrient recovered in soils is essential for maintaining high yields and soil fertility to sustain agriculture production over the long term. Smart nutrient management program has considerable potential to enhance growth of rice in developing countries in the next decade.

Acknowledgements

The authors would like to thanks Dr. Tariq Aziz, project director of University of Agriculture Faisalabad, Pakistan and an anonymous review for valuable constructive criticism that improved this manuscript.

Conflict of interest

There is no conflict of interest.

Author details

Rai Mukkram Ali Tahir*[1], Noor-us-Sabah[1], Muhammad Afzal[2], Ghulam Sarwar[1] and Ijaz Rasool Noorka[1]

1 Department of Soil and Environmental Sciences, College of Agriculture, University of Sargodha, Pakistan

2 Department of Agricultural Entomology, College of Agriculture, University of Sargodha, Pakistan

*Address all correspondence to: rai786@gmail.com

References

[1] Mukkram AT, Sarwar G, Afzal M, Sabah NU, Muhammad S. Physiochemical transformation of waste of super phosphate industry into new generation silicate fertilizer and its uses for rice growth under saline environment. Pakistan Journal of Botany. 2018;**50**(1):117-122. DOI: https://www.pakbs.org/pjbot/papers/1531140570

[2] Ludlow MM, Muchlow RR. A critical evaluation of traits for improving crop yields in water limited environments. Advances in Agronomy. 1990;**43**:107-153. DOI: 10.1016/S0065-2113(08)60477-0

[3] Fageria NK, Baligar VC, Jones CA. Growth and Mineral Nutrition of Field Crops. 3rd ed. New York: CRC Press; 2010. ISBN: 9781439816950

[4] Fageria NK, Moreira A. The role of mineral nutrition on root growth of crop plants. In: Advances in Agronomy. Vol. 110. Burlington: Academic Press; 2011. pp. 251-331. DOI: 10.1016/B978-0-12-385531-2.00004-9

[5] Fageria NK. Nutrient management for improving upland rice productivity and sustainability. Communications in Soil Science and Plant Analysis. 2001;**32**:2603-2629

[6] Fageria NK. The Use of Nutrients in Crop Plants. Boca Raton, FL: CRC Press; 2011

[7] RRI (International Rice Research Institute). Rice Facts. Los Banos, Phillipines: IRRI; 1993

[8] Fageria NK, Moreira A. Yield and yield components of upland rice as influenced by nitrogen sources. Journal of Plant Nutrition. 2010;**32**:361-370. DOI: 10.1080/01904167.2011.536878

[9] Fageria NK, Baligar VC. Methodology for evaluation of low land rice genotype for nitrogen use efficiency. Journal of Plant Nutrition. 2003;**26**:1315-1333. DOI: 10.1081/PLN-120020373

[10] Gruhn P, Goletti F, Yudelman M. Integrated Nutrient Management, Soil Fertility and Sustainable Agriculture: Current Issues and Future Challenges. Food, Agriculture and Environment Discussion Paper 32. Washington, DC, USA: International Food policy Research Institute. 2000. p. 38. DOI: https://ageconsearch.umn.edu/bitstream/16236/1/dp000032.pdf. ISBN: 0-89629-638-5

[11] Mahajan A, Bhagat RM, Gupta RD. Integrated nutrient management in sustainable rice-wheat cropping system for food security in India. SAARC Journal of Agriculture. 2008;**6**(2): 1-16. Available from: http://www.saarcagri.net/

[12] Savant NK, De Datta SK. Nitrogen transformation in wetland rice soils. Advances in Agronomy. 1982;**35**: 241-302. DOI: 10.1007/BF00750514

[13] Hasegawa H. High yielding rice cultivars perform best even at reduced nitrogen fertilizer rate. Crop Science. 2003;**43**:921-926. DOI: 10.7685/j.issn.1000-2030.2008.01.013

[14] Kyi M, Kumudar WM, Win KK, Yamakawa T. Combined effect of 0rganic manures and inorganic fertilizers on the growth and yield of hybrid rice [palethwe-1]. American Journal of Plant Sciences. 2017;**8**: 1022-1042. DOI: 10.4314/wajae.v16i1.55866

[15] Qadir MA, Ahmad M, Ihsan NU, Asrar A, Waseem R. EDTA chelated micronutrients for growth of rice, onion and lemon plants and enhancing their result yields. International Journal of Chemical Sciences.

2014;**12**(4):1153-1160. DOI: 10.4236/ajps.2015.67094

[16] Fageria NK, Stone LF. Micronutrient deficiency problems in South America. In: Alloway BJ, editor. Micronutrient Deficiencies in Global Crop Production. New York: Springer; 2008. pp. 247-268. DOI: 10.1007/978-1-4020-6860-7_10

Section 4

Rice Utilization, Processing and Marketing

Chapter 11

Formulation and Nutritional Assessment of Recipes En Route for Awareness of Coarse Rice Consumption

A. Arun

Abstract

Rice is one of the oldest domestic crops being consumed by the human kind centuries back; it has been popular and a common staple food for people across the world. Rice grain has undergone various genetic evolutions for centuries. Presently, a variety of crops are grown with altered genetic nature, induced with highly toxic fertilizers and chemical insecticides. The quantity of harvest is more concerned than the quality of the grains. Being the staple food for millions of people in Asian countries, it has been portrayed as the main cause for many deadly diseases. The chapter involves in creating awareness in rice consumption through formulating healthy common recipes formulated and infused with coarse rice varieties.

Keywords: coarse rice, healthy food, nutrition, recipes

1. Introduction

Globally, the food pattern is getting worse and adversely influenced by the ostentatious behavior among the younger generations. Among the feature that manipulates healthy eating, unavailability of healthy food commodities acts as the prime fact [1]. Consumption of food in its natural state has become extinct; artificial treatments of food commodities are widespread. Food in its natural state is more nutritious and wholesome with all essential requirements; once been treated or fragmented, the food is malformed from its actual value. The food commodities are chemically treated for a long shelf life due to commercialized and industrial motives.

Rice is a staple food being commonly consumed across many countries and been portrayed as the best branded crop with good genetic map [2]. During the crop year 2016–2017, worldwide 161.1 million hectares of land was under rice cultivation, and India was the prime producer of rice to harvest 44.5 million hectares [3]. Rice varieties are categorized upon the aspects like texture, shape, length, aroma, and color.

The nutritious compilation of rice is highly influenced by the environmental facts like soil, irrigation, and climatic conditions. Rice is a composition of carbohydrate-starch (75–80%) [4]; vitamins such as thiamine, niacin, and riboflavin [5]; highly digestible protein; and water (12%). About 302 proteins are recognized in proteome of the rice seeds that symbolize 252 gene products [6]. Besides this, rice is also rich in other micronutrients and minerals such as copper, iron, manganese, calcium, phosphorous, and zinc [7].

Rice is nutritious with low fat and sodium that prevents high blood pressure and cardiovascular diseases. It is gluten free and rich in fibers that fight against cancerous cells [8]. Rice being a staple food consumed in its natural state possesses all values of importance; being chemically treated, polished, and packed, it lose its significance.

The rice paddy undergoes a number of refining processes to turn out to be white rice; at the intermediate stage, the rice with outer bran and germ portion called brown rice is being processed, that is, removal of outer bran layer which is called the white rice. The brown rice contains five times more fiber content than the white rice [9]. The presence of outer bran layer enhances these rice varieties in its nutritional value, color, taste, texture, and flavor. Coarse rice varieties are nutty and chewy in nature; also takes an extra cooking time compared to white rice. Coarse rice is categorized upon the size of the rice and the thickness of bran layer present in the rice that determines its color [10].

Brown rice, wild rice, or coarse rice is consumed more in the rural regions of the country. These unpolished rice or brown rice accompanied with an outer covering called bran makes the rice wholesome with fibers, minerals, and vitamins [11]. At the side of the nutritious facts, consumption of these coarse rice varieties helps the farmers in developing their economic background. The farmer can recover about 7.5 kg of brown rice from 10 kg of paddy, whereas by further milling, the cost of white rice is increased by 50–60%, while the quantity is reduced to 6.5 kg [12]. Production cost of coarse rice is comparatively lesser than the white rice [13]; the demand for the white rice has reduced the production of brown rice and made it a premium product; thus, an increase in consumption of these coarse rice varieties may bring more varieties of brown rice into the market [14].

In an urge to search for healthy diet, many traditional rice varieties have been restored and brought into regular consumption beside using only for family functions and religious rituals. These rice varieties have also taken a prime role in the diet pattern of people in urban regions. All over the world, many traditional rice varieties have been recovered; these restored rice varieties can withstand climatic changes and can give a good yield with natural organic fertilizers and insecticides.

In the study area, many traditional rice varieties has been restored and brought into practice some such popular varieties used in the formulation of recipes are mapillai samba arisi (groom samba rice), karuppu kavuni arisi (black cow rice), kattu yanam arisi (wild rice), moongil arisi (bamboo rice), and red glutinous rice (sigappu kavuni arisi) [15].

These coarse rice are regaining their lost values as they are organic in nature with no additives or undergone any chemical treatments. The study enumerates some simple and common sweet recipes of these rice varieties.

2. Materials and methods

The study involves formulation of recipes with the available coarse rice varieties commonly found in southern part of India. There exist no special criteria in selection of recipes; besides the only factor considered is the recipe to be common and consumable at any time with no restriction to age.

The common sweet recipes consumed in India are formulated and infused with these coarse rice varieties. The ingredients (**Table 1**) required for the study are procured from the nearby organic stores, and the recipes' formulation and preparation were done.

Ingredients	Kheer (g)	Pan cake	Puttu (g)	Kolukattai (g)	Laddoo	Kesari
Rice	200	200 g	200	200	200 g	200 g
Jaggery	200	100 g	–	100	–	–
Coconut	50	25 g	200	100	–	–
Ghee	50	50 ml	–	–	50 ml	50 ml
Cashew nut	20	10 g	–	10	25 g	20 g
Sultanas	10	–	–	–	–	20 g
Cardamom	5	5 g	5		5 g	5 g
Palm sugar	25	50 g	100	50	200 g	200 g

Note: the table lists out the ingredients used in formulating recipes.

Table 1.
Ingredients used in the recipes.

The mise-en-place is done before the initiation of cooking. The rice varieties, coconut, and palm sugar are grated or blended as per the requirement for the recipes. The jaggery is made into thick syrup and filtered to remove the residues and used for cooking, which also reduces the cooking time.

2.1 Methods of preparation

The recipes have their own methods of cooking. The formulated recipes are preferred to be prepared with organic natural sweeteners like palm sugar or jaggery than refined white sugar. Palm sugar and jaggery are natural sugars, and enhances the recipes with their natural flavor, color, and taste. They are rich in micronutrients and vitamins [16].

The methods of preparations of the recipes are listed below:

- Mapillai samba rice kheer:
    - Cook the rice and coarsely smash and keep aside.
    - In a large pan, add ghee, roast sultanas, and cashew nut.
    - Add required amount of water and let it boil.
    - Add natural sweeteners, palm sugar and jaggery syrup, and the smashed rice.
    - Reduce the flame and then add coconut milk slowly.
    - Mix well and finally add a pinch of salt and cardamom powder.

Serve the recipe hot or refrigerated; it can be diluted with coconut milk if been thick. Garnish with roasted nuts and sultanas.

- Black kavuni rice pan cake:
    - The rice has to be soaked for 2 h at least.
    - Grind well the rice into a thick paste and dilute it with thick coconut milk for a rich taste.

- Add chopped nuts, grated coconut, and a pinch of salt and cardamom powder.
- Let the batter rest for a few minutes and then finally add the jaggery syrup and palm sugar.
- Then start making pan cakes and serve hot.

- Wild rice puttu (**Figure 1**):
    - The wild rice is cleaned, washed, shade dried, and grinded in a blender.
    - Sieve the flour and add the cardamom powder.
    - Slightly sprinkle a small amount of salt water mix well and add grated coconut gradually and steam cook the mixture for 20 min.
    - Remove it from the heat, add finely powdered palm sugar, garnish with grated fresh coconut, and serve hot.

- Bamboo rice kolukattai (modak):
    - Stuffing:
        1. In a hard bottomed pan, add ghee and then grated coconut, and saute well for 5 min.
        2. Gradually add the jaggery syrup and palm sugar to the coconut and then the chopped cashew nuts.
        3. Mix well the mixture as the water reduces remove from flame and keep aside and let it cool.
    - Dough:
        1. Sieve the bamboo rice flour with salt and add hot water to it gradually and mix well to turn into soft dough. Let the dough cools.
        2. Put the soft dough in a modak die with the stuffing in the center and keep it aside.
        3. Arrange the stuffed dough in a steamer and cook the preparation for 30 min and serve hot.

- Navara rice laddoo (**Figure 2**):
    - Roast the rice into a golden brown color and grind it well.
    - Sieve the flour, add a pinch of salt and cardamom powder, and mix it well.
    - Take a thick bottomed pan heat and add ghee.
    - Gradually add the rice flour mixture and then powdered palm sugar and mix well.

DOI: http://dx.doi.org/10.5772/intechopen.85628

  - Stir well till the flour and palm sugar mixes well; add roasted cashew nut to enrich the taste; add ghee if required.
  - Once the mixture blends well, remove from the fire and make small balls (lemon size) out of the mixture.

- Rose matta rice kesari (**Figure 3**):
  - Roast the rice, let it cool, and grind coarsely.
  - In a large pan, roast the sultanas and cashew nut into golden brown color then add water and let it boil.
  - Add palm sugar and boil the mixture for few minutes.
  - To the hot water mixture, add the coarsely grinded rice slowly without forming lumps. Let the mixture cook well and add ghee gradually.
  - Garnish with roasted cashew nut and sultanas. Serve hot.

Figure 1.
Wild rice puttu.

Figure 2.
Navara rice laddoo.

Figure 3.
Rose motta rice kesari.

The formulations of sweet recipes are more concerned for health concepts; besides the ingredients used in recipes, the healthy cooking methods like boiling and steaming are more considered in the preparations.

The recipes formulated can also be infused with other varieties of coarse rice with same methods of preparation, while there is a small alter in cooking time depending upon the variety of coarse rice used in the recipe.

2.2 Assessment of nutritional composition of the recipes

The recipes were analyzed for the importance of nutritional composition; the prepared samples of rice recipes are sent to the Food Science Laboratory. The recipe samples were homogenized with electric blender and dried in electric oven. These dried rice recipe samples are blended into powder and used for the examination. The moisture content, fat, protein, and carbohydrates energy were calculated, and further analysis was carried out to exhibit the presence of minerals and micronutrients like iron, calcium, zinc, and vitamins.

Moisture content and total ash of the rice recipes were determined according to the standard methods of IS 1011 [17]. Protein was determined by IS 7219 [18], the micro-Kjeldahl method that determined the amount of nitrogen in the sample, which was subsequently multiplied by a factor of 6.25. Total fat in the sample was determined using Soxhlet extraction apparatus [Model Soxtherm Automatic by AOAC] [19]. The remaining percentage represented carbohydrates by difference (Food and Agriculture Organization of the United Nations) [20]. Energy was calculated from fat, carbohydrate, and protein contents using Atwater's conversion factors (Food and Agriculture Organization of the United Nations) [21].

Minerals such as zinc, calcium, and iron were determined by following the method of AOAC (20th edition) 999.11:2016 [22], AOAC (20th edition) 927.02 [23], and AOAC (20th edition) 975.03 [24], respectively. Vitamins like A, E, C, B_1, B_2, and B_3 were determined by following the method AOAC. Vitamin A and E were analyzed by AOAC method of 992.06 [25]. Vitamin C was estimated by AOAC [26] method 967.21, whereas vitamin B_1 by AOAC [27] method of SMPR 2015.002. Vitamin B_2 and B_3 contents were estimated by AOAC [28] method of SMPR 2015.003 and AOAC [29] method of SMPR 2015.004, respectively.

3. Results and discussions

The study was carried out to bring out awareness about the consumption of rice in its actual state than chemically treated or polished. Being the staple food in many countries, our meal revolves around rice or rice products; consumption of rice is an unavoidable fact but the white rice can be replaced with these coarse rice varieties. Many countries have initiated creating awareness programs toward the consumption of these coarse rice varieties. They explain the positive effects of consuming coarse rice by organizing exhibitions on rice varieties, conducting cooking competitions, food festivals, posting in social media, etc.

The study enlists few common sweet recipes formulated and infused with these traditional and wholesome coarse rice varieties. The sweet recipes are too common and consumed all over the country (**Table 2**).

3.1 Mapillai samba rice kheer

The kheer is a sweet dish popularly called payasam in south India, which is consumed in all parts of the country with different names [30]. Formulation of kheer with mapillai samba rice and coconut milk makes it more healthy and nutritious. The recipe is rich in iron (8.872 mg) and zinc (3.58 mg) that helps to maintain a healthy nervous system as the rice itself is rich in these micronutrients [31] Vitamins like vitamin C (61.8 mg), niacin (2.06 mg), and riboflavin (1.48 mg) present in the recipes make the recipe nutritious.

3.2 Black kavuni rice pan cake

Black kavuni rice has a high esteem value being consumed by royal peoples; the rice is energy rich a good source of protein [32] and all phyto-chemical contents [33], and the nutrients are naturally present in the plants. The rice is a good source of vitamin B12 and has a high glycemic index [34] that reduces the accumulation of hepatic fat and helps to recover liver damages [35]. The pan cake recipe is rich in calcium (340.25 mg), phosphorus (1030 mg), and vitamin B12 cobalamin (12 mg). The recipe with black rice is rich in antioxidants that help to fight cancer and cardiovascular diseases.

3.3 Wild rice puttu

Puttu is a steamed dish commonly consumed in southern part of India. It can be prepared with any rice variety or millets [36]. The recipe is rich in magnesium (385 mg) and vitamin B5 pathogenic acid (9.43 mg) and a good source of fibers (12.405 g). The nutrients and fiber present in the recipe help in treatment for constipation and diabetic patients [37].

3.4 Bamboo rice kolukattai (modak)

Bamboo rice is one among the staple food of the tribal peoples [38]. The recipes made out of bamboo rice are healthier and nutritive as it is recommended for the pregnant ladies to compensate diet needs [39]. The recipe is rich in calcium (232.5 mg), potassium (247 mg), and magnesium (286 mg); it helps in fighting cholesterol and maintains blood pressure.

Parameter	Composition (mean value) of nutrients in the recipes					
	Kheer	Pan cake	Puttu	Kolukattai	Laddoo	Kesari
Moisture (g)	122.93 ± 0.12	52.19 ± 0.41	40.6 ± 0.77	30.99 ± 1.01	15.075 ± 0.12	50.532 ± 0.66
Protein (g)	21.1 ± 0.19	18.48 ± 0.14	25.91 ± 0.12	20.02 ± 0.69	21.41 ± 0.94	20.35 ± 0.47
Total fat (g)	81.215 ± 0.11	66.25 ± 0.01	83.21 ± 2.1	46.84 ± 0.31	63.035 ± 1.45	60.69 ± 1.98
Minerals (g)	4.05 ± 0.45	4.335 ± 0.02	5.07 ± 0.4	3.84 ± 0.04	5.67 ± 0.17	5.55 ± 0.1
Fiber (g)	7.665 ± 0.4	8.535 ± 0.51	12.405 ± 0.94	8.53 ± 0.44	4.73 ± 0.004	17.665 ± 0.14
Carbohydrates (g)	383.54 ± 0.32	357.21 ± 0.12	280.805 ± 0.31	314.78 ± 0.11	361.08 ± 0.14	359.965 ± 7.1
Energy (kcal)	2341.4 ± 4.11	2060.55 ± 0.98	1920.45 ± 2.1	1742.1 ± 0.6	2018.45 ± 1.33	1988.65 ± 0.78
Calcium (g)	260.25 ± 0.14	340.25 ± 1.4	281.5 ± 2.39	232.5 ± 1.89	489 ± 3.95	286.5 ± 1.30
Phosphorus (mg)	579 ± 1.9	492 ± 1.77	652 ± 4.0	567 ± 7.4	528.5 ± 1.21	506 ± 2.14
Iron (mg)	8.872 ± 0.9	5.251 ± 0.12	4.83 ± 0.89	6.221 ± 0.2	3.0825 ± 0.14	1.63 ± 0.03

Note: the table exhibits the mean result of nutritional values from laboratory analysis report.

Table 2.
Nutritious assessment of the formulated recipes.

3.5 Navara rice laddoo

Navara rice is a popular variety of rice used in Kerala Ayurvedic medicines, which help in treating arthritis and many neuro disorders [40]. The recipe has protein (21.41 g), vitamin C (73.1 mg), and niacin (4.201 mg). This rice variety is more helpful to reduce pain due to arthritis, strengthen bones, and promote the growth of red blood cells [41].

3.6 Rose matta rice kesari

Rose matta rice is a popular coarse rice variety consumed commonly in Kerala. This variety of rice is a good source of fiber, calcium, magnesium, and vitamin A [39]. The kesari recipe formulated with this rice is enhanced by the rice's natural pink color. The recipe is nutritious with a high content of calcium (286.5 mg) and fiber (17.665 g).

The recipes formulated can be also infused and altered with other coarse rice varieties; the nutritive value is more dependent upon the coarse rice variety used. On an overall analysis, the recipes formulated with coarse rice are rich in vitamins like vitamin B and vitamin C, calcium, phosphorous, magnesium, protein, and fiber.

4. Conclusion

Consumption of polished white rice has been considered as an esteemed status among the people at urban regions. The study has emphasized awareness among the common people in bringing out the importance, nutritive value, and health benefits of the available coarse rice varieties. Thus, the study suggests the inclusion of coarse rice or rice as wholesome grain and can be included in the daily recipes with innovative formulations; inclusion of this wholesome grain reduces the risk of many deadly diseases.

Conflict of interest

The author declares no conflict of interest.

Ethical clearance

The author declares no ethical clearance.

Author details

A. Arun
School of Hotel and Catering Management, Vels Institute of Science Technology and Advanced Studies, Chennai, India

*Address all correspondence to: arunarticle2016@gmail.com

References

[1] Imamura F, Micha R, et al. Dietary quality among men and women in 187 countries in 1990 and 2010: A systematic assessment. The Lancet Global Health. 2015;**3**(3):Pe132-Pe142. DOI: 10.1016/S2214-109X(14)70381-X

[2] Sasaki T, Matsumoto T, Gojobori T. The genome sequence and structure of rice chromosome. Nature. 2002;**420**:312-316

[3] Greenfield M. World Rice Acreage 2008-2017 [Internet]. Cited 2018 November 27. Statista.com. Available from: https://www.statista.com/statistics/271969/world-rice-acreage-since-2008/

[4] Verma DK, Srivastav PP. Proximate composition, mineral content and fatty acids analyses of aromatic and non-aromatic indian rice. Rice Science. 2017;**24**(1):21-31. DOI: 10.1016/j.rsci.2016.05.005

[5] Rohman A, Helmiyati S, Hapsari M, Setyaningrum DL. Rice in health and nutrition. International Food Research Journal. 2014;**21**(1):13-24

[6] Yang Y, Dai L, Xia H, Zhu K, Liu H, Chen K. Protein profile of rice (Oryza sativa) seeds. Genetics and Molecular Biology. 2013;**36**(1):87-92

[7] Oko AO, Ubi BE, Efisue AA. A comparative study on local and newly introduced rice varieties in Ebonyi State of Nigeria based on selected agronomic characteristics. International Journal of Agriculture and Forestry. 2012;**2**(1): 11-17. DOI: 10.5923/j.ijaf.20120201.03

[8] Koga M, Kiso Y, Kusumi I, Toyomaki A. PT739. Rice is nice: Benefits of rice consumption for brain health. The International Journal of Neuropsychopharmacology. 2016;**19**(Suppl 1):69. DOI: 10.1093/ijnp/pyw044.739

[9] Kondo K, Morino K, Nishio Y, Ishikado A, Arima H, Nakao K, et al. Fiber-rich diet with brown rice improves endothelial function in type 2 diabetes mellitus: A randomized controlled trial. PLoS One. 2017;**12**(6):e0179869. DOI: 10.1371/journal.pone.0179869

[10] Longnecker D. Types of Brown Rice. Livestrong.com [Internet]. Cited on 2018 December 28. Available from: https://www.livestrong.com/article/547899-types-of-brown-rice/

[11] NDTV Food Desk. 9 Health Benefits of Switching to Brown Rice. 2018 July 03. [Internet]. Cited 2018 December 27. Available from: https://food.ndtv.com/food-drinks/9-health-benefits-of-switching-to-brown-rice-1730477

[12] The Manila Times. Philrice Urges More Brown Rice Consumption Updated 2016 September 02 [Internet]. Cited on 2018 December 31. Available from: https://www.manilatimes.net/philrice-urges-more-brown-rice-consumption/283468/

[13] Loida EM, Reforma MG. An exploratory market study of brown rice as a health food in the Philippines. Journal of Agricultural Science. 2010;**55**(1):19-30

[14] Emme Rose Santiagudo. DA Promotes Brown Rice Consumption. Updated 2018 November 9 [Internet]. Cited on 2018 December 31. Available from: https://thedailyguardian.net/local-news/da-promotes-brown-rice-consumption/

[15] Hema Menon. Traditional rice varieties [Internet]. Cited on 2018 December 24. Available from: http://hemamenan.blogspot.com/2017/06/paarambriya-arisi-vagaigal.html

[16] Srikaeo K, Thongta R. Effects of sugarcane, palm sugar, coconut sugar

and sorbitol on starch digestibility and physicochemical properties of wheat based foods. International Food Research Journal. 2015;**22**(3):923-929

[17] IS 1011 (2002): Biscuits [FAD 16: Foodgrains, Starches and Ready to Eat Foods. Indian Standard. Biscuits–Specification. (Fourth Revision). Bureau of Indian Standards. Printed at New India Printing Press: Khurja, India

[18] IS 7219 (1975): Edible Groundnut Flour (Expeller Pressed) [FAD 16: Foodgrains, Starches and Ready to Eat Foods]. Indian Standard. Specification For Edible Groundnut Flour (Expeller Pressed) (First Revision). Bureau Of Indian Standards. Printed at New India Printing Press: Khurja, India

[19] AOAC, editor. Official Methods of Analysis. 17th ed. Gaithersburg MD: AOAC International; 2000. Official Method 920.39

[20] Food Energy–Methods of Analysis and Conversion Factors. Food and Agriculture Organization of the United Nations—ISSN0254-4725. Food and Nutrition Paper 77. 2003:12

[21] Food Energy–Methods of Analysis and Conversion Factors. Food and Agriculture Organization of the United Nation—ISSN0254-4725. Food and Nutrition Paper 77. 2003:23

[22] AOAC, editor. Official Methods of Analysis. 20th ed. Virgina, USA: AOAC International; 2002. 999.11:2016

[23] AOAC, editor. Official Methods of Analysis. 20th ed. Virgina, USA: AOAC International; 2002. 927.02

[24] AOAC, editor. Official Methods of Analysis. 20th ed. Virgina, USA: AOAC International; 2002. 975.03

[25] AOAC, editor. Official Methods of Analysis. 17th ed. Gaithersburg MD: AOAC International; 2000. Official Method 992.06

[26] AOAC, editor. Official Methods of Analysis. 17th ed. Gaithersburg MD: AOAC International; 2000. Official Method 967.21

[27] AOAC. Standard Method Performance Requirements SM (SMPRs) for Total Vitamin B1 (Thiamin) in Infant and Adult/Pediatric Nutritional Formula. Rockville, MD USA: AOAC International; 2015 SMPR 2015.002

[28] AOAC SMPR. Standard Method Performance Requirements SM (SMPRs) for Total Vitamin B2 (Riboflavin) in Infant and Adult/ Pediatric Nutritional Formula. Rockville, MD USA: AOAC International; 2015. SMPR 2015.003

[29] AOAC SMPR. Standard Method Performance Requirements SM (SMPRs) for Total Vitamin B3 (Niacin) in Infant and Adult/Pediatric Nutritional Formula. Rockville, MD USA: AOAC International; 2015. SMPR 2015.004

[30] Barela SR, Shelke RR. studies on acceptability, chemical composition and cost structure of kheer prepared from cow milk blended with coconut milk. International Journal of Current Microbiology and Applied Sciences. 2017;**6**(11):2527-2532. ISSN: 2319-7706

[31] Aravind Bharadwaj D. Mappillai Samba: The Forgotten Treasure of Health. Updated on 2018 March 1 [Internet] Cited on 2019 January 1. Available from: https://www.practo.com/healthfeed/mappillai-samba-the-forgotten-treasure-of-health-32307/post

[32] Kalaivani R et al. A study on medicinal properties of traditional rice karung kavuni and neutraceutical formulation. International Journal of Food and Nutritional Science. 2018;**5**(1):86-90

[33] Rajeswari K, Kiranmai E, Sharma S. Preliminary screening of

phytochemical content of different cereals grown in different states of India. International Journal of Chemical Studies. 2018;**6**(2):2272-2275

[34] Mattoo S. Black Rice Is the Latest Super Food. Here's Why! Updated on 2016 January 05 [Internet] Cited on 2019 January 03. Available from: https://timesofindia.indiatimes.com/life-style/health-fitness/diet/Black-rice-is-the-latest-superfood-Heres-why/articleshow/50439583.cms

[35] Jang H-H, Park M-Y, Kim H-W, et al. Black rice (*Oryza sativa* L.) extract attenuates hepatic steatosis in C57BL/6 J mice fed a high-fat diet via fatty acid oxidation. Nutrition and Metabolism. 2012;**9**(1):1. DOI: 10.1186/1743-7075-9-27

[36] Sajeev J, Sasikumar S, Eagappan K. Standardization and organoleptic evaluation of South Indian recipes incorporated with *Tribulus terrestris* (L) fruit powder. International Journal Of Food And Nutritional Sciences. 2015;**4**(4)

[37] Umar MA, Ugonor R, Akin-Osanaiye CB, Kolawole SA. Evaluation of nutritional value of wild rice from Kaduna State, Central Nigeria. International Journal of Scientific and Technology Research. 2013;**2**(7):140-147

[38] Kiruba S, Jeeva S, Das SSM, Kannan D. Bamboo seeds as a means to sustenance of the indigenous community. Indian Journal of Traditional Knowledge. 2007;**6**(1):199-203

[39] Selvarajan S et al. Influence of rice varieties in diabetics among indian population—A review. European Journal of Pharmaceutical and Medical Research. 2016;**3**(8):184-188. Available from: http://www.ejpmr.com/admin/assets/article_issue/1469856627.pdf

[40] Boominathan M, Bakiyalakshmi SV. Analysis of bioactive compounds in navara (njavara) rice by gcms. International Journal of Recent Scientific Research. 2016;**7**(11): 14307-14311. Available from: http://www.recentscientific.com

[41] Chaudhari PR, Tamrakar N, et al. Rice nutritional and medicinal properties: A review article. Journal of Pharmacognosy and Phytochemistry. 2018;**7**(2):150-156. Available from: http://www.phytojournal.com/archives/2018/vol7issue2/PartC/7-1-256-103.pdf

Section 5

Modern Toolboxes for Crop Protection

Chapter 12

Toxic Potential of *Bacillus thuringiensis*: An Overview

David Fernández-Chapa, Jesica Ramírez-Villalobos and Luis Galán-Wong

Abstract

The toxins of *Bacillus thuringiensis* (Bt) have shown great potential in the control of harmful insects affecting human health and agriculture, used as the main biological agent for the formulation of bioinsecticides due to its specificity to target different insects' orders. This has led Bt-based products to become the best-selling biological insecticides in the world since the genes encoding insecticidal proteins have been successfully used in novel insecticidal formulation, genetically engineered (GE) crops, and development of transgenic rice that produce insecticidal toxins derived from *Bacillus thuringiensis*. It has been proven that insecticidal activity of Bt protein crystals can prolong their toxicity in shelf life or field under specific conditions, and this can improve the use of special strains and formulations to control insect vectors of diseases. Bt toxins have shown well-documented toxicity against lepidopterans, coleopterans, hemipterans, dipterans, nematodes, Rhabditida and human cancer cells of various origins. These crystal toxins may be responsible for other novel biological properties suggesting a pluripotential nature with different specificities.

Keywords: *Bacillus thuringiensis*, Cry toxins, bioinsecticide, resistance, Dulmage

1. Introduction

In the modern era, *Bt* was isolated for the first time in Japan by the bacteriologist Ishiwata Shigetane in 1901, and it was considered the microorganism responsible for the disease of the silkworm sotto *Bombyx mori*. The author named it *Bacillus sotto*, which means soft and flaccid, in reference to the appearance of the infected larvae. He noted that young bacterial cultures were not pathogenic to larval insects; in contrast old cultures that suffered sporulation were highly toxic. However, the first valid description was until 1911, when the German scientist, Ernst Berliner, isolated it from diseased larvae of the flower moth *Anagasta kuehniella*. He named it *Bacillus thuringiensis*, which derives from Thuringia, the German town where moths were found [1].

Bacillus thuringiensis is a ubiquitous gram-positive, rod-shaped soil bacterium, that has been isolated worldwide from a great diversity of ecosystems including soil, water, dead insects, dust from silos, leaves from deciduous trees, diverse conifers, and insectivorous mammals [2–4], known by its ability to produce crystalline inclusions during sporulation (Cry toxins) which contain insecticidal proteins called δ-endotoxin. Crystalline inclusions from Bt are showing well-documented toxicity to a wide variety of insect pests, such as Lepidoptera, Coleoptera, and Diptera [5], hemipterans, as other biological activities such as molluscicidal, nematicide (human and animal parasites, and free living; Rhabditida), acaricide and even against human cancer cells [2, 6–10].

Bt toxins have been applied to the environment since 1933 and began to be used commercially in France in 1938, and by 1958 their use had spread to the United States. From the 1980s Bt becomes a pesticide of global interest [11].

Bt crystal and secreted soluble toxins are highly specific for their hosts and have gained worldwide importance as an alternative to chemical insecticides. Bt toxins have been considered as the most successful bioinsecticide during the last century. Currently, it consists of more than 98 (424 million USD) of formulated sprayable bacterial pesticides [12] and is the most common environmental-friendly insecticide used and is the basis of over 90% of the pesticides available in the market today [13].

2. Bioinsecticide activity of *Bacillus thuringiensis* proteins

The main difference between *Bacillus thuringiensis* and other closely related bacillus is the formation, during the sporulation process, of one or more crystalline bodies of a protein nature adjacent to the spore. Some of these parasporal crystals known as δ-endotoxins (Cry and Cyt) confer the pathogenic capacity against larvae of different orders of insects, mostly Lepidoptera, Diptera, Coleoptera and in some cases against species of other phyla [14]. By synthesizing parasporal crystalline inclusion during sporulation, the bacterium can ensure its survival, since a dead insect can provide sufficient nutrients that allow the spores to germinate [15].

Bt strains synthesize crystal (Cry) and cytolytic (Cyt) toxins (also known as δ-endotoxins), at the onset of sporulation and during the stationary growth phase as parasporal crystalline inclusions. Additionally, Bt isolates can also synthesize other insecticidal proteins during the vegetative growth phase; these are subsequently secreted into the culture medium, the vegetative insecticidal proteins (Vip) [5, 16], and the secreted insecticidal proteins (Sip) [17].

This part refers to the nomenclature first used for Cry genes, on the next part of the page it explains the nomenclature currently used for Bacillus thuringiensis genes [18] (**Table 1**).

However, this nomenclature was not ideal, since the new toxins had to be tested against an increasing number of insects so that the toxin and the gene could be

Main classes	Order	Cry toxins
Group 1	Lepidoptera	Cry1, Cry9, and Cry15
Group 2	Lepidopteran and dipterous	Cry2
Group 3	Coleoptera	Cry3, Cry7, and Cry8
Group 4	Diptera	Cry4, Cry10, Cry11, Cry16, Cry17, Cry19, and Cry20
Group 5	Lepidoptera and Coleoptera	Cry1I
Group 6	Nematodes	Cry6

Table 1.
Classification of Cry toxins according to their insect host specificities proposed by Crickmore et al. [18].

named; that was when the *Bacillus thuringiensis* Toxin Nomenclature Committee was created in 1993 and proposed a new classification system [18], which consists of giving the new toxin a four-rank name depending on its degree of pairwise amino acid identity to previously named toxins, using Arabic numbers for the first and fourth rank and uppercase and lowercase letters for the second and third ranks, respectively, for example, Vip1 and Vip2 if they share less than 45% pairwise identity, Vip3A and Vip3C if they share less than 78% pairwise identity, Vip3Aa and Vip3Ab if they share less than 95% pairwise identity, and Vip3Aa1 and Vip3Aa2 if they share more than 95% pairwise identity [19].

Based on the amino acid sequences, there are 75 families of Cry proteins, with 800 different *Cry* genes [20], while the Cyt proteins consist of three families with 38 genes [21].

Cry proteins have been reported to be toxic to Lepidoptera, Coleoptera, Hymenoptera, Hemiptera, Diptera, Orthoptera, and Mallophaga and also against nematodes, mites, and Protozoa (**Figure 1**) [22]. Some toxins have an expanded

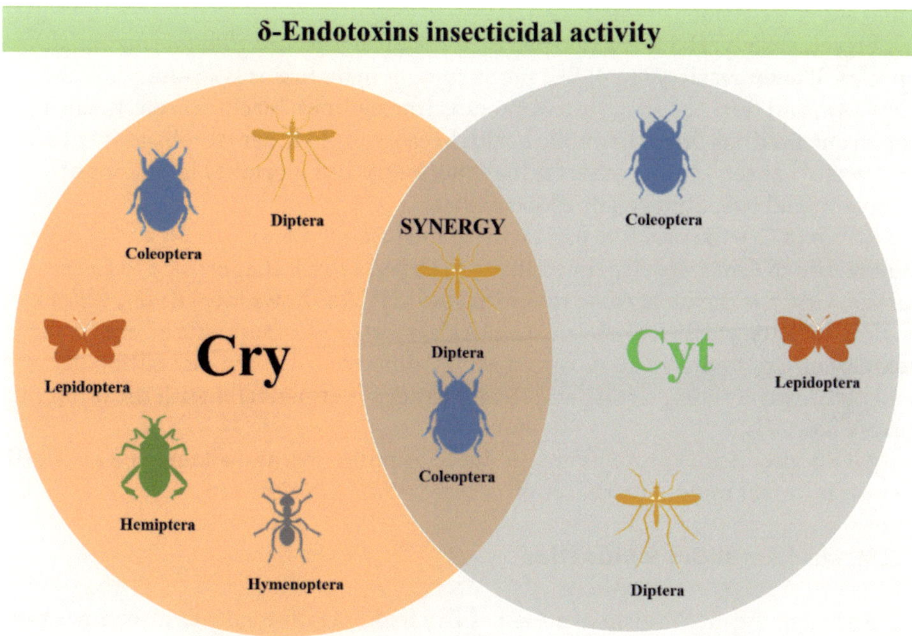

Diptera	Cry1(A-C), Cry2(A-B), Cry4(A-C), Cry10(A), Cry11(A-B), Cry16(A), Cry17(A), Cry19(A-C), Cry20(A), Cry21(A-H), Cry24(A-C), Cry25(A), Cry27(A), Cry29(A-B), Cry30(A-G), Cry32(B-D), Cry39(A), Cry40(A-D), Cry44(A), Cry47(A), Cry48(A),Cry49(A), Cry52(A-B), Cyt1(A-D), Cyt2(A-D), Cyt3(A)
Coleoptera	Cry1(A-I), Cry2(A), Cry3(A-C), Cry7(A), Cry8(A-G), Cry9(D), Cry14(A), Cry18(A-C), Cry22(A-B), Cry23(A), Cry26(A), Cry28(A), Cry34(A-B), Cry35(A-B), Cry36(A), Cry37(A), Cry38(A), Cry43(A-C), Cry55(A), Cyt1(A), Cyt2(C)
Lepidoptera	Cry1(A-K), Cry2(A-B), Cry7(B), Cry8(D) Cry9(A-C,E), Cry15(A), Cry19(A-C), Cry20(A-B), Cry22(A), Cry32(A), Cry51(A), Cry54(A-B), Cry59(A-B), Cyt2(B)
Hemiptera	Cry3(A), Cry5(A), Cry22(A)
Hymenoptera	Cry2(A), Cry3(A), Cry11(A)

Figure 1.
Insecticidal activity of Cry and Cyt δ-endotoxins against the orders Diptera, Coleoptera, Lepidoptera, Hemiptera, and Hymenoptera [15, 21, 23].

spectrum of action to two or more order or phylum [10]. For example, Cry1B is one of those that present a remarkable activity against larvae of Lepidoptera, Diptera, and Coleoptera. So, the combination of toxins present in a strain will define its spectrum of action [4].

In contrast, Cyt toxins have predominant activity against dipterous; however, they have toxic activity against some lepidopteran and coleopteran [24]; in addition, some Cyt toxins are able to establish synergy for insecticidal activity with other Bt proteins such as Cry or Vip3 and to reduce the resistance levels of Cry proteins toward some insect species of the Coleoptera and Diptera orders (**Figure 1**). The Cyt1Aa toxin from *Bacillus thuringiensis* var. israelensis is active against *Chrysomela scripta* and *Culex quinquefasciatus* and can prevent the development of resistance to the proteins Cry3Aa, Cry4, and Cry11Aa [14].

2.1 Bti toxins

Bacillus thuringiensis subsp. israelensis (Bti) was first isolated from a water pond in the Negev desert [25] and was the very first strain described for having insecticidal activity outside Lepidoptera.

Bti serovariety, H-14, is a subspecies of the diversified *Bacillus thuringiensis* species. The serovariety H-14, Bti, produces four main toxins (Cry4Aa, Cry4Ba, Cry11Aa, and Cyt1Aa) specific to dipterans (mosquitoes, blackflies, etc.) which represent a serious threat to public health because of their hematophagous nature and vector capacity responsible for high morbidity and mortality in billions of people spread over almost half of the planet.

Bti toxin Cry4Ba is active primarily against *Anopheles* and *Aedes* and shows no toxicity to *Culex* species, in contrast to Cry4Aa toxin that is toxic to *Culex* larvae. Cry11 is the most toxic to Aedes, and Cyt1Aa shows low (*Aedes*, *Culex*) to non-toxicity at all (*Anopheles*). Cyt1Aa has a strong synergistic effect on the toxicity of Cry toxins in all mosquitoes. In addition to its own mosquitocidal and cytotoxic activity, Cyt1A was shown to act synergistically with the other Bti toxins [26, 27].

All Bti insecticidal proteins are produced as protoxins, and all must be activated in vivo by insect midgut proteases prior insecticidal activity.

2.2 Mechanism of Cry toxin action

Although the mechanism of action of Cry toxins against various insects has been widely investigated, there are still many controversies. Therefore, there are currently different models in the literature that seek to explain it [28].

The sequential union model is known as the classical mechanism. It has been detailed in studies with the Cry1Ab protein in *Manduca sexta*. It postulates that the toxic properties come from crystalline inclusions produced during the sporulation of *Bt*. The crystals and their subunits are inert protoxins and are not biologically active, and their mode of action can be plotted as follows: the δ-endotoxins are ingested, the crystals are solubilized by the alkaline pH of the intestine, the inactive protoxins are digested by proteases of the midgut which produces an active toxin of about 60–70 kDa resistant to proteases, and then the Cry toxins come into contact with the N-aminopeptidase receptors and cadherin on the surface of the membrane. The affinity between toxins and certain types of receptors results in proteolysis of the Cry protein that causes structural changes in the chains and forms oligomers that function as "pre-pores." The N-aminopeptidase receptor anchors the pre-pore in the lipid bilayer, pore formation affects integrity of the membrane,

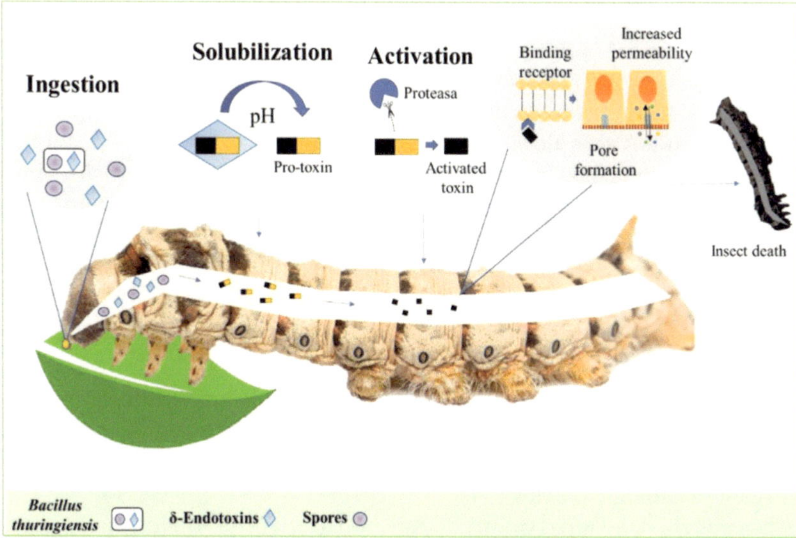

Figure 2.
Mechanism of action of Cry proteins according to the sequential binding model.

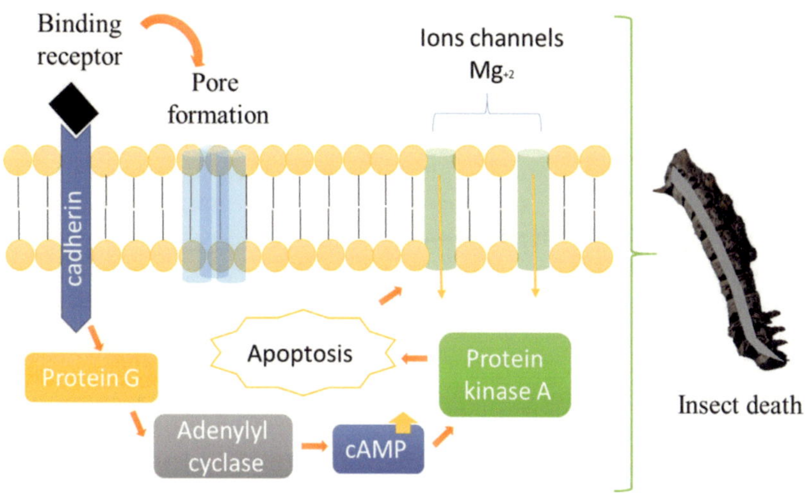

Figure 3.
Mechanism of action of Cry proteins according to the signaling pathway model.

and electrophysiological evidence and biochemistry suggest that the pores cause an osmotic imbalance that causes cell death and lysis; the intestine is paralyzed, the insect stops feeding, and there is diarrhea, total paralysis, and finally death (**Figure 2**) [1, 29].

The second proposed mechanism called signaling pathway model has similarities with the previous model; however, in this other causes for cell death are assigned. According to this theory, Cry proteins affect the cell in two ways: first by the formation of pores in the membrane, as mentioned in the sequential binding model and, second, by the production of successive reactions that alter the cellular metabolism. According to this hypothesis, Cry toxins bind to cadherin receptors, which stimulate heterotrimeric G protein and adenylyl cyclase with an increase in cAMP production. The cAMP activates the protein kinase A, which stimulates apoptosis with an activation of the Mg^{2+} channels in the plasma membrane. The opening of these channels causes an abnormal movement of the ions in the cytosol, stimulating the process of apoptosis (**Figure 3**) [1, 3, 30].

The germination of the spores also contributes to the death of insect, since the vegetative cells can replicate within the host's hemolymph and cause septicemia; however, the δ-endotoxins alone are sufficient to kill some insect species if they are produced in high doses. This feature has been exploited by expressing the delta endotoxin genes in bacteria that better adapt to a particular environment, as well as its expression in genetically modified plants [31, 32].

3. Howard T. Dulmage's methods and contributions on *Bt*

Howard T. Dulmage was a microbiologist who established his line research in the study of pathogenic bacteria insects [33] and is considered one of the most important pioneers in the development of technologies for the implementation of *Bt* as a control agent of biological pests [34].

Working at the US Department of Agriculture, at the Agricultural Research Service (USDA-ARS), in Brownsville, Texas, Howard T. Dulmage from the pink worm, *Pectinophora gossypiellia*, a strain of *Bacillus thuringiensis* variety kurstaki, in 1969, which is 200 times more active in laboratory tests against the pink bollworm, *Pectinophora gossypiella*; the tobacco budworm, *Heliothis virescens*; and the cabbage looper, *Trichoplusia ni* [21], higher than that of the known strains, which is marketed as "Dipel" by the company United States of America [35–37].

Strain HD-1 is one of the best-studied strains, since it is characterized by the carrying of a variety of Cry anti-Lepidoptera genes, *Cry1Aa*, *Cry1Ab*, *Cry1Ac*, *Cry2Aa*, *Cry2Ab*, and *Cry1Ia*, and since its discovery, the outlook for *Bt*-based products has expanded and is still the most commercial success of microbial control of pest [4, 38].

H. Dulmage sets up the basis for the fermentation and formulation procedures of Bt culture extracts for their commercialization [39] and were among the most important pioneers in the development of technologies for the implementation of Bt as a biological pest control agent. He established diverse methodologies for mass production product formulation and power standardization [40].

At the beginning of the 1970s, two great advances were obtained by Dulmage, the first was based on the recovery of the spore-crystal complex by means of precipitation with lactose-acetone to produce powders and wettables, which was rapidly developed and adapted in the industry. The second was the adoption of a standardized system to calibrate the potency of the different preparations of *Bt* and the establishment of international toxicity units (ITU)/mg, which allowed the comparison of the different products developed [41, 42]. The equation proposed by Dulmage [43] is the following:

$$\text{Test extract potency (ITU/mg)} = \frac{\text{standard } LC_{50}[\text{standard potency (ITU/mg)}]}{\text{test extract } LC_{50}} \quad (1)$$

Dulmage established better bioassay methods to assess the effectiveness of powders [37, 44].

In 1984, Dulmage participated in the establishment of a bioassay protocol for toxicity assessment of *Bacillus thuringiensis* var. israelenensis powders. This protocol differs from the one previously suggested by WHO in the Guidelines for Bti Production regarding the follow aspects:

- Specifies a standard cup for larval exposure to Bti extracts.
- Establishes a number of 20 larvae per cup and three replications for the concentrations assayed.
- If a minimum of six extract concentrations is tested, a repetition of the assay is required.
- A computational probit analysis is required for evaluating the toxicity as LC50.
- A mortality or pupation higher than 5% in the control invalidates the bioassay.

Additionally, the study suggested a variability coefficient of less than 20% for each repetition. Dulmage, together with a team of colleagues, tested the validity of this protocol and suggested some considerations for the management of the reference standard strains and for the establishment of new ones.

3.1 Howard T. Dulmage's fermentation extracts

From 1970 to 1988, Dulmage established the largest Bt collection in the Americas, and he collected more than 800 isolates that were named using his HD code, belonging to 21 serovarieties. From these 800 isolates, 17 belonged to the H-14 serovariety, corresponding to Bti. He conducted a series of fermentation experiments with Bt in order to optimize the production and to assess the effectiveness of powder; hundreds of fermentation extracts were generated, and some of them were donated by the US Department of Agriculture in 1989 to the International Collection of Entomopathogenic Bacillus of the Faculty of Biological Sciences of the University of Nuevo León, Mexico, which has approximately 4000 stored fermentation extracts of which 3000 of them correspond to HD strains, and currently extracts are found in the form of dry powder, with different times of storage [38].

3.2 Bti strain collection

In the 1970s, Dulmage continued to the control of disease-transmitting mosquito larvae using lepidopteran-active isolates having some reported dipteran activity. When Dulmage became aware of the discovery of a new Bt subspecies capable of attacking dipteran larvae, especially simuliids (*Bacillus thuringiensis* subsp. *israelensis*) (Bti), he quickly perceived the great value of this discovery, because of the possibility to control dangerous human disease vectors, and began to be involved in studies on Bti as dipteran biocontrol agent.

One of the greatest contributions of Dulmage to Bti research was the compilation of a protocol guide for Bt H-14 serovariety local production. This guide was an extension of the procedures developed by him for the production, formulation, and standardization of lepidopteran-specific serovarieties. These guidelines were presented and discussed in the informal consultation on local H-14 Bt production, in Geneva, Switzerland, in October 1982. The 128-page booklet was prepared by Dr. Dulmage, at the request of the Scientific Working Group on biological control of vectors of the Special Program for Research and Training in Tropical Diseases of the World Health Organization, and was published in 1983 [45].

In 1985, Dulmage and a research group proved the tested strain was Bti HD-968-S-1983, which resulted to be 4.74 times more potent than the standard use

(IPS-78); the potency assigned to it was 4740 ± 398 ITU/mg. They recommended the use of this strain as the potency reference standard for comparison with any Bti formulation.

Twenty samples of the strain HD-500 and HD-567 of Bti fermentation extracts from the collection of Dulmage et al. [44] recovered by lactose-acetone coprecipitation during the period from 1978 to 1983 maintained their residual toxic activity against the mosquito *Aedes aegypti*. All extracts evaluated presented toxicity at the highest tested doses (1000 ppm), and two of the stored extracts (3260 and 3501) showed LD50 of 0.12 and 1.16 ppm, respectively [40].

Bti protein crystals from fermentation extracts showed persistence of toxic activity of fermentation extracts after more than three decades. This opens the possibility of improving the use of special strains and improved formulations to control insect vectors of diseases.

4. New Cry toxins

Despite the success of the application of Bt crystal proteins for the biological control of pests, at present it is still necessary to identify new Cry toxins with greater toxicity; this approach is considered one of the best ways to counter the potential resistance evolved by insects as well as in developing products against a wider spectrum of insect pests. Traditionally, Bt isolates were screened for their insecticidal spectrum by the time-consuming and laborious insect bioassays [22, 46]. Since only a limited number of cry genes have been used for insect control either in sprays or transgenic crops so far, novel insecticidal genes are required [31].

The most common technique used to predict toxicity is the polymerase chain reaction (PCR), through the identification of new cry genes [47], but high-throughput sequencing technology has also been used in the discovery of toxins [20]. Seventy-two antigenic groups (serovariety) have been distinguished for *Bacillus thuringiensis* [48]. Crickmore et al. [19] have designed an especial database for Bt toxins with links to information on host insects, based on the last update (www.lifesci.sussex.ac.uk/Home/Neil_Crickmore/Bt/). About 952 toxin genes, encoding different entomopathogenic proteinaceous toxins, have been identified and characterized in the Bt strains isolated all around the world; however, only a small proportion of these proteins are highly toxic and therefore used in the production of bioinsecticides. This can be accomplished by either finding new wild-type strains or engineering Cry proteins with enhanced activity or altered insecticidal spectrum by swapping domains and site-directed mutagenesis; nevertheless a thorough knowledge of Cry protein structure and binding interactions with target receptors is a must [49].

Additionally, the construction of *Bt* DNA libraries in *Escherichia coli*, followed by screening by Western blotting or a hybridization-based method, or the development of DNA libraries in an acrystaliferous mutant of *Bt* followed by microscopic observation and/or SDS-polyacrylamide gel (SDS-PAGE) detection of expressed genes has also been used to detect novel Cry protein genes [44].

Moreover, a combination of genomics, transcriptomics, proteomics, and metabolomics could be used to study *Bt* toxin proteins with different characteristics and activities [21]. However, due to the interaction between different toxins produced by a strain in insect midgut, bioassays provide complementary and necessary characterization information. Due to the money, time, and material costs associated with insect rearing and time-consuming characteristics of insect bioassays, cell-based assays have been employed for toxicity characterization of *Bt* strains or toxins [50].

Furthermore, recent studies have confirmed more new potentials of different Bt strains. These new features are including plant growth promotion [51], bioremediation of heavy metals and other chemicals [1, 52], anticancer activities [53], polymer production [54], and antagonistic effects against plant and animal pathogenic microorganisms [55].

5. *Bacillus thuringiensis* development on rice crops

Genetically engineered or transgenic crops producing Cry proteins from *Bacillus thuringiensis* are key management tools against several important insect pests. GE plants expressing Bt insecticidal proteins selectively target insect pests while having little impact on beneficial insects. Bt toxins have been widely adopted worldwide; it was calculated that over 100 million hectares of crops contained Bt genes by 2017 [56].

Bt crops produce either a single toxin or more than one Bt toxin; these are called pyramided crops. Bt pyramided crops delay evolution of resistance to target pests, insects resistant to one toxin are killed by other toxins in the pyramid [57, 58]. Nevertheless, pyramided Bt crops are vulnerable to the development of cross-resistance. The use of Bt pyramids and the simultaneous planting of non-Bt crops are the main strategies applied to produce susceptible pest insects (known as the "refuge strategy") [59].

Rice is a primary food source for more than half of the world's population making it one of the most fundamental crops. Since 1989 multiple insect-resistant genetically engineered (IRGE) rice lines expressing *Bacillus thuringiensis* insecticidal proteins had been developed [60], controlling lepidopteran pests. There are four major lepidopteran pest rice such as the rice stem borers *Chilo suppressalis* (Walker) (Lepidoptera: Crambidae), *Scirpophaga incertulas* (Walker) (Lepidoptera: Crambidae), *Cnaphalocrocis medinalis* (family Crambidae), and *Sesamia inferens* (Walker) (Lepidoptera: Noctuidae) [61].

Bt rice lines resistant to rice lepidopteran pests mainly express Cry1Aa, Cry1Ab, Cry1Ac, Cry1B, Cry1C, Cry1Ca1, Cry2A, and Cry9C proteins [61–63].

Since Cry1Ab was first introduced into a japonica rice variety, many Bt genes have been found, and only a few of them were selected for developing transgenic crops [60]. Because deploying two or more Bt genes in one rice variety can delay the emergence of pest resistance [64, 74], scientists started to develop Bt hybrid rice lines with Cry1Ab/Cry1Ac into various rice plants which have both high grain yield and good grain quality [65].

Some advantages of expressing fusion proteins like Cry1Ab/Cry1Ac and Cry1Ab/Vip3A are the equalization of the expression level of the two proteins, trait integration in different crops, and highly efficient expression strains [66]. Studies on Cry1Ab/Cry1Ac fusion protein have demonstrated great effectiveness significantly reducing the incidence of *Chilo suppressalis* [67, 72].

Other *B. thuringiensis* proteins that present high affinity are Cry9Aa and Vip3Aa. These two proteins bind specifically to brush border membrane vesicles of the Asiatic rice borer *Chilo suppressalis*, which do not share binding sites [68]. Cry9Aa and Vip3Aa toxins have shown potent toxic synergy based on a specific interaction between them against *C. suppressalis* larvae with a synergism factor (SF) value of 10.6-fold [68].

The rice water weevil (*Lissorhoptrus oryzophilus* Kuschel) is another of the most destructive insect pests of cultivated rice (*Oryza sativa*) in the United States [69, 70]. This pest causes low yields in rice by damaging the roots from larval feeding in the submerged root zone [71].

Some of the strategies to control this insect pest are the use of pyrethroids, which are toxic to aquatic organisms [72], synthetic insecticides, and weed control around fields to reduce habitat for rice water weevil adults. *Bacillus thuringiensis* spp. *galleriae* (Btg) have proven to be an environmentally friendly alternative against rice water weevil larvae. Studies indicate that Btg granular formulation has biological activity against this target pest and performs as well as the pyrethroids insecticides [73], showing promising potential for rice water weevil control.

6. Resistance to *Bacillus thuringiensis*

Bt insecticides consist of several types of insecticidal crystal proteins; hence, the development of insecticidal pesticide resistance is difficult or slow [47]. However, resistance has already been observed in laboratory, and the first case was a population of Indian meal moths, *Plodia interpunctella*, in 1985, and since then different insect species have been reported to be resistant to one or more *Bt* toxins under laboratory conditions. However, the situation in the field remains very different. To date, the only natural populations that have really developed resistance following *Bt*-based treatments have been populations of diamondback moth, *Plutella xylostella* [27, 74].

The use of transgenic plants has greatly increased the selection pressure on target pest populations and is likely to become much more acute in natural conditions if *Bt* use in agriculture and for human health applications spreads or in cases of the nonrational use of large-scale transgenic crops expressing *cry* genes [32].

In agriculture worldwide, repeated applications of Bt sprays and widespread adoption of Bt crops (transgenic crops protected from insects by the expression of *Cry* and/or *Vip3* genes) have led to resistance [75, 76].

Field populations of *Diabrotica virgifera* have shown resistance to eCry3.1Ab maize and cross-resistance among Cry3Bb1, mCry3A, and eCry3.1Ab, which are the Bt toxins most commercialized for management of western corn rootworm [77].

Resistance to Cry toxins can be developed by mutations in the insect pests that affect any of the steps of the mode of action of Cry toxins [78]. "Field populations" refers to insects on the field, since the conditions are distinct in vitro, can be developed by different mechanisms, such as altered activation of Cry toxins by midgut proteases sequestering the toxin by glycolipid moieties or esterases, by inducing an elevated immune response, and by alteration resulting in reduced binding to insect gut membrane; among all these mechanisms of resistance, the most common mechanism of toxin resistance is the reduction in toxin binding to midgut cells, which in different resistant insect species include mutations in Cry toxin receptors such as cadherin (CAD)-like proteins, alkaline phosphatase (ALP), or aminopeptidase N (APN) or mutations in the ABCC2 transporter [78].

The emergence of resistant insects is a problem that both *Bt* sprays and plant products are likely to face in the future [32]. Several strategies, such as the use of spatial or temporal refugia, high or ultrahigh doses, and gene pyramiding to express two toxins, or two insect control approaches, such that the possibility of evolution of resistance to two toxins/approaches, independent of each other, is greatly diminished, can be a promising approach to prolong the efficacy of products based on *Bt* [36, 46].

There are different methods to counteract the resistance of insects to Bt toxins, for example, assisted mutation with UV light; the combination of Bt toxins with other toxins, such as *Bacillus sphaericus* proteins; and formulations with plant extracts.

Nevertheless, a new method has been used to combat resistance to Bt toxins, the phage-assisted continuous evolution (PACE), which rapidly evolves Bt toxins to bind a new receptor with high affinity and specificity, expressed on the surface of insect midgut cells. The PACE system enhances the insecticidal activity against both sensitive and Bt-resistant insect larvae up to 335-fold, through more than 500 generations of mutation, selection, and replication to bind a new receptor [23]. Collectively, these methods establish an approach to overcoming Bt toxin resistance.

7. Formulations based on *Bacillus thuringiensis*

The production of toxic proteins has given *Bacillus thuringiensis* enormous interest in its inclusion in phytosanitary formulations. The efficiency of products based on *Bt* depends on the type of formulation, as well as various environmental factors. Formulation depends on the persistence of toxicity and the choice of application method; other important factors are UV radiation, agitation, sedimentation, water quality, contaminants, pH, temperature, susceptibility of insects, and competition with other microorganisms [79].

The wide variety of formulations based on spores and crystals intended for being ingested by the white insect are the result of many years of research. The development of a large variety of spore-crystal complex matrices allows for improvements, such as increased toxic activity, increased palatability to insects, or longer storage times. These matrices use chemical, vegetable, or animal products, which are constituted in such a way that they favor contact between crystals and insects, without harming humans or the environment [80].

Proper formulation can help to overcome several of the factors that limit or reduce its larvicidal activity and improve control performance by enabling greater contact with target larvae, ensuring stability under storage and field conditions, providing a variety of application options, and increasing the ease of handling. There are several types of formulations, among the most used are:

Powder (DP)

- Formulated by sorption of an active ingredient on finely ground mineral powder (talc, clay, etc.).
- Particle size of 50–100 μm.
- Powders can be applied directly to the target, either mechanically or manually.
- The inert ingredients for this formulation are anticaking agents, ultraviolet protectors, and adhesive materials to improve adsorption.
- Concentration of the active ingredient (organism) in the powder is usually 10%.

Granules (GR)

- Granular particles are larger and heavier than powder formulations.
- Particle size coarse of 100–1000 μm for granules and 100–600 μm for microgranules.
- Made of mineral materials (kaolin, attapulgite, silica, starch, polymers, dry fertilizers, and residues of ground plants) [81].
- Concentration of the active ingredient (organisms) in granules ranges from 5 to 20%.
- Once applied, the granules slowly release their active ingredient.
- Some granules require soil moisture to release their active ingredient [3, 82].

Wettable powders (WP)

- Finely ground dry formulations that will be applied after suspension in water.
- Produced by mixing an active ingredient with surfactants, wetting and dispersing agents, and inert fillers, followed by milling.
- Particle size approximately 5 μm.
- Long storage stability, good miscibility with water, and convenient application with conventional spray equipment [83].

Water dispersible granules (WG)

- Designed to be suspended in water.
- The granules break to form a uniform suspension similar to that formed by a wettable powder.
- Compared to powdered products, these WGs are relatively dust-free and with good storage stability.
- The products contain a wetting agent and dispersing agent similar to those used in wettable powders, but the dispersing agent is usually at a higher concentration.

The emulsions

- Consist of liquid droplets dispersed in another immiscible liquid.
- Size of the droplets in the dispersed phase varies from 0.1 to 10 μm.
- The emulsion can be oil in water (EW), which is a normal emulsion, or water in oil (EO), an inverted emulsion. Both products are designed to be mixed with water before use.

The suspension concentrate (SC)

- A mixture of a finely ground solid active ingredient dispersed in a liquid phase, usually water.
- The solid particles do not dissolve in the liquid phase, so that the mixture needs to be stirred before application to keep the particles evenly distributed.
- The composition of the suspension concentrate is complex and contains wetting/dispersing agents, thickening agents, antifoaming agents, etc., to ensure the required stability.
- They are produced by a wet milling process.
- Particle size distribution of 1–10 μm.

Oil dispersions (OD)

- Dispersions of solid active ingredients in a nonaqueous liquid intended for dilution before use.
- The nonaqueous liquid is more often an oil (vegetable oil).
- Oil dispersion provides several important characteristics, such as the ability to supply water-sensitive active ingredients and the ability to use an adjuvant fluid instead of water that can increase and extend pest control.

Capsule suspension (CS)

- Stable suspension of microencapsulated active ingredient in an aqueous continuous phase.
- Intended for dilution with water before use.
- The bioagent as an active ingredient is encapsulated in capsules (coating) made of gelatin, starch, cellulose, and other polymers.
- Protected from extreme environmental conditions (UV radiation, rain, temperature, etc.).
- Residual stability increases due to slow (controlled) release.
- The most frequently applied encapsulation method uses the principle of interfacial polymerization.

The extension of pesticide formulations containing Bt will depend essentially on our capacity to improve the performance of the products used [83]. Therefore, biotechnology companies have the task of providing not only formulations adapted to certain crops and insect pests, but also, they must look for and produce bioinsecticides based on the new high-potency strains originating from the agroecosystems where they are going to apply. It is expected that the new products that appear in the market will provide a spectrum of higher activity that will impact on a greater number of pests in other crops and can help develop sustainable agriculture [80].

8. Bioinsecticides based on *Bt*

Worldwide, the use of biopesticides increases 16% annually, which represents approximately 8% of the pesticide trade in the world [12]. The formulations derived from natural materials such as bacteria, animals, plants, or minerals offer a powerful tool to create a new generation of sustainable products [84]. About 90% of microbial biopesticides are derived from a single entomopathogenic species *Bacillus thuringiensis* [85].

Bt-based bioinsecticides are classified into first-line products up to the fourth generation: (1) They are made up of spores and crystals, have several drawbacks, since they present a narrow range of activity when more than one pest insect is present, have little persistence in the field to solar radiation, and do not reach insects that attack roots or internal parts of the plant. (2) They contain spores and toxins of strains as an active ingredient with the introduction of genes from other

Bt variety	Susceptible insects	δ-Endotoxin	Producer company
kurstaki	Lepidoptera	Cry1Aa, Cry1Ab, Cry1Ac, Cry2Aa, and Cry2Ab	Abbott-Dupont and Certis
aizawai	Lepidoptera	Cry1Aa, Cry1Ab, Cry1Ba, Cry1Ca, and Cry1Da	Abbott-Dupont and Kenogard
san diego	Coleoptera	Cry3Aa	Mycogen
tenebrionis	Coleoptera	Cry3Aa	Thermo Trilogy, Columbia MD, Certis Mycogen, and Novo Nordisk
israelensis	Diptera	Cry4A, Cry4B, Cry11A, and Cyt1Aa	Abbott-Dupont, Novo Nordisk, and Certis
galleriae	Coleoptera	Cry8Da	Phyllom BioProducts

Table 2.
Varieties of Bt used as bioinsecticides, susceptible insects, expressing δ-endotoxin, and companies that produce it.

strains, which is very useful to improve the action against the insect, generating a synergism, as well as diminishing the possibilities of resistance. (3) They contain recombinant bacteria, especially Pseudomonas fluorescens or *Clavibacter xyli* subsp. *cynodontis*, which are able to reach plant tissues and grow in the rhizosphere. (4) They constitute protein chimeras [86].

The varieties of *Bt* used commercially for the production of bioinsecticides for the control of Lepidoptera are *kurstaki* and *aizawai*, for Coleoptera the *san diego*, *tenebrionis* and *galleriae* are used, and for the control of dipteros, the *israelensis* is the most used (**Table 2**) [48, 74, 78, 87].

9. Applications

More than a century after its discovery, *Bt* has become an important tool for the management of insect pests, whether in the agricultural sector or in the fight against vectors of diseases. Since then the spectrum of its applications has been increasing and is no longer limited to its initial function. It has become evident that the potential of *Bt* would transcend the biological control of insects, and recent studies analyze new properties for this old acquaintance [88].

These new environmental features include the toxicity against nematodes, mites, and ticks, antagonistic effects against plant and animal pathogenic bacteria and fungi, plant growth-promoting rhizobacteria (PGPR) activities, bioremediation of different heavy metals and other pollutants, biosynthesis of metal nanoparticles, production of polyhydroxyalkanoate biopolymer, and anticancer activities (due to parasporins) [51–53].

Toxicity against nematodes with several classes of Cry toxin (Cry5, Cry6, Cry13, Cry14, Cry21, and Cry55) is well established. In addition to these Cry proteins, thuringiensin, chitinase, and a metalloproteinase from *Bt* are also toxic to nematodes [89]. In contrast, the information about the effect of *Bt* on mites is rare, and a few in vitro and in vivo studies have reported the acaricidal activity of some *Bt* strains. In a study conducted by Dunstand et al. [90], the in vitro acaricidal activity was reported to be caused by the strain GP532 of *Bt* on the mite *Psoroptes cuniculi*. Histological alterations caused by *Bt* on this mite included the presence of dilated intercellular spaces in the basal membrane, membrane detachment of the peritrophic matrix, and morphological alterations in columnar cells of the intestine.

Cry proteins synthesized by *Bt* do not show any antifungal activity. However, some *Bt* strains produce antifungal compounds, including cell wall-degrading enzymes, lipopeptide fengycin [21]. In a study conducted by Shrestha et al. [91], *Bt* strain C25 was antagonistic to *Sclerotinia minor* and *Sclerotinia sclerotiorum*, and it was found that the strain was capable of inhibited mycelial growth, suppressed sclerotia formation, and germination. On the other hand, strain C25 showed high activities of various cell wall-degrading enzymes such as proteases, β-1,3-glucanase, and chitins.

Some strains of *Bt* colonize plant roots and have plant growth-promoting characteristics. Many *Bt* strains produce some metabolites which enhance plant growth at abiotic stress conditions. These compounds include 1-aminocyclopropane-1-carboxylic acid (ACC) deaminase, indole-3-acetic acid (IAA), proline, phosphate solubilization enzymes, and siderophore production [92].

Different strains of *Bt* have been shown to produce many potential factors that could be of great interest in the biocontrol of phytopathogenic bacteria [55]. *Bt* produces bacteriocins, chitinases, acyl homoserine lactone lactonase, and zwittermicin, which collectively elicit detrimental effects on insect hosts and target bacteria; although the role of *Bt* bacteriocins in nature is enigmatic, it is possible that they assist in pathogenesis by attacking competing endosymbiotic or opportunistic bacteria, thereby facilitating propagation of this entomopathogen in the hemolymph of susceptible insects [93].

Parasporins are a heterogenous group of Cry proteins produced by noninsecticidal *Bt* strains that specifically act on human cancer cells without affecting normal ones, and it has been reported that Cry proteins, such as Cry31A, Cry41A, Cry45A, Cry46A, Cry63A, and Cry64A, present anticancer activity when digested with proteases [53].

10. Advantages and disadvantages

The biopesticide based on bacteria is probably the most used and is cheaper than the other methods of bioregulation of pests [94]. Almost 90% of the microbial biopesticides that are commercially available are *Bt* derivatives [95]. Among the advantages and disadvantages of using Bt as a bioinsecticide are the following [34] (**Table 3**):

Advantages	Disadvantages
Performance: although each kilogram is more expensive, only a few grams per hectare are needed compared to 4 kg of chemical insecticides	Application with difficulty
High toxicity: a small amount is needed to kill pests	It is not easy to produce it
Specificity: it only kills the target organism	Little diffusion and acceptance by producers
It does not produce infections: it is demonstrated that an infected larva does not harm other insects, animals, or even humans	Its quality could not be controlled. Sometimes it works, and sometimes it does not
Limited time of permanence in the environment: after 3 or 4 weeks of application, traces of the bioinsecticide are no longer found	Variability in insect resistance
Few cases of resistance: there are few cases reported, and only in extraordinary conditions there are certain degrees of resistance	Location. Its use may be limited to faunas of a certain region

Table 3.
Advantages and disadvantages of bioinsecticides based on Bt.

11. Conclusion

Bacillus thuringiensis has undoubtedly been the most successful microbial agent for biological insect control of all time. However, different authors have warned of the generation of insect resistant to Bt-derived products, as well as genetically modified plants.

During the last two decades, new methods have been widely used on Bt to overcome resistance to insects, and it is expected that this advancing trend will be well continued in the future, including the search for new toxins and strains with increased toxic activity and the development of new biopesticides and technologies to maintain the success of this bioinsecticide which is a great challenge to overcome.

Nowadays there exist different lines of research that seek to use *Bt* in different applications, such as anticancer activity, promotion of plant growth; nematicide, antifungal and bactericidal activity among others. To achieve the implementation of these new features, it is necessary to know more about the biochemical and physiological pathways, as well as the mode of action of the new features. Such properties will undoubtedly lead to explore novel Bt strains with more potent insecticide activities or novel features which will enhance the implementation of these strains in other medical, agronomical, and industrial avenues. At the same time, technological development is necessary to allow new products to become a reality.

Conflict of interest

The authors declare that they have no conflicts of interest.

Author details

David Fernández-Chapa, Jesica Ramírez-Villalobos and Luis Galán-Wong*
Instituto de Biotecnología, Facultad de Ciencias Biológicas, Universidad Autónoma de Nuevo León, Ciudad Universitaria, San Nicolas de los Garza, NL, Mexico

*Address all correspondence to: ljgw55@gmail.com

References

[1] Melo AL, Soccol VT, Soccol CR. *Bacillus thuringiensis*: Mechanism of action, resistance, and new applications: A review. Critical Reviews in Biotechnology. 2016;**36**:317-326

[2] Höfte H, Whiteley HR. Insecticidal crystal proteins of *Bacillus thuringiensis*. Microbiological Reviews. 1989;**53**:242-255

[3] Knowles BH, Dow JAT. The crystal delta-endotoxins of *Bacillus thuringiensis*—Models for their mechanism of action on the insect gut. Bioessays. 1993;**15**:469-476

[4] Roh JY, Choi JY, Li MS, Jin BR, Je YH. *Bacillus thuringiensis* as a specific, safe, and effective tool for insect pest control. International Journal of Microbiology Biotechnology. 2017;**17**:547

[5] Estruch JJ, Warren GW, Mullins MA, Nye GJ, Craig JA, Koziel MG. Vip3A, a novel *Bacillus thuringiensis* vegetative insecticidal protein with a wide spectrum of activities against lepidopteran insects. Proceedings of the National Academy of Sciences. 1996;**93**:5389-5394

[6] Abd El-Ghany A, Abd El-Ghany N. Molluscicidal activity of *Bacillus thuringiensis* strains against *Biomphalaria alexandrina* snails. Beni-Suef University Journal of Basic and Applied Sciences. 2017;**6**:391-393

[7] Chougule NP, Bonning BC. Toxins for transgenic resistance to hemipteran pests. Toxins. 2012;**4**:405-429

[8] Ohba M, Mizuki E, Uemori A. Parasporin, a new anticancer protein group from *Bacillus thuringiensis*. Anticancer Research. 2009;**29**:427-433

[9] Schnepf E, Crickmore N, van Rie J, Lereclus D, Baum J, Feitelson J, et al. *Bacillus thuringiensis* and its pesticidal crystal proteins. Microbiology and Molecular Biology Reviews. 1998;**62**:775-806

[10] Van Frankenhuyzen K. Cross-order and cross-phylum activity of *Bacillus thuringiensis* pesticidal proteins. Journal of Invertebrate Pathology. 2013;**114**:76-85

[11] Ochoa G, Arrivillaga J. *Bacillus thuringiensis*: Avances y perspectivas en el control biológico de *Aedes aegypti*. Boletín de Malariología y Salud Ambiental. 2009;**49**:181-191

[12] Lacey LA, Frutos R, Kaya HK, Vail P. Insect pathogens as biological control agents: Back to the future. Journal of Invertebrate Pathology. 2015;**132**:1-41

[13] Cherif A, Ettoumi B, Raddadi N, Daffonchio D, Boudabous A. Genomic diversity and relationship of *Bacillus thuringiensis* and *Bacillus cereus* by multi-REP-PCR fingerprinting. Canadian Journal of Microbiology. 2007;**53**:343-350

[14] Palma L, Muñoz D, Berry C, Murillo J, Caballero P. *Bacillus thuringiensis* toxins: An overview of their biocidal activity. Toxins. 2014;**6**:3296-3325

[15] De Maagd R. How *Bacillus thuringiensis* has evolved specific toxins to colonize the insect world. Trends in Genetics. 2001;**17**:193-199

[16] Donovan WP, Engleman JT, Donovan JC, Baum JA, Bunkers GJ, Chi DJ, et al. Discovery and characterization of Sip1A: A novel secreted protein from *Bacillus thuringiensis* with activity against coleopteran larvae. Applied Microbiology and Biotechnology. 2006;**72**:713-719

[17] Warren GW, Koziel MG, Mullins MA, Nye GJ, Carr B, Desai NM. Auxiliary Proteins for Enhancing the

Insecticidal Activity of Pesticidal Proteins. U.S. Patent 5,770,696; 1998

[18] Crickmore N, Zeigler DR, Feitelson J, Schnepf E, van Rie J, Lereclus D, et al. Revision of the nomenclature for the *Bacillus thuringiensis* pesticidal crystal proteins. Microbiology and Molecular Biology Reviews. 1998;**62**:807-813

[19] Crickmore N, Zeigler DR, Schnepf E, van Rie J, Lereclus D, Baum J, et al. *Bacillus thuringiensis* toxin nomenclature. Microbiology and Molecular Biology Reviews. 1998;**62**:807-813

[20] Sajid M, Geng C, Li M, Wang Y, Liu H, Zheng J, et al. Whole genomic analysis of *Bacillus thuringiensis* revealing partial genes as a source of novel Cry toxins. Applied Microbiology and Biotechnology. 2018;**84**:e00277-e00218

[21] Jouzani GS, Valijanian E, Sharafi R. *Bacillus thuringiensis*: A successful insecticide with new environmental features and tidings. Applied Microbiology and Biotechnology. 2017;**101**:2691-2711

[22] Ye W, Zhu L, Liu Y, Crickmore N, Peng D, Ruan L, et al. Mining new crystal protein genes from *Bacillus thuringiensis* on the basis of mixed plasmid-enriched genome sequencing and a computational pipeline. Applied Microbiology and Biotechnology. 2012;**78**:4795-4801

[23] Badran A, Guzov V, Huai Q, Kemp M, Vishwanath P, Kain W. Continuous evolution of *Bacillus thuringiensis* toxins overcomes insect resistance. Nature. 2016;**533**:58-63

[24] Soberón M, López-Díaz JA, Bravo A. *Bacillus thuringiensis*: A protein fold conserved in several pathogenic microorganisms. Peptides. 2013;**41**:87-93

[25] Goldberg LJ, Margalit J. A bacterial spore demonstrating rapid larvicidal activity against *Anopheles sergentii*, *Uranotaenia unguiculata*, *Culex univitattus*, *Aedes aegypti* and *Culex pipiens*. Mosquito News. 1977;**37**:355-358

[26] Canton PE, Reyes EZ, de Escudero IR, Bravo A, Soberon M. Binding of *Bacillus thuringiensis subsp. israelensis* Cry4Ba to Cyt1Aa has an important role in synergism. Peptides. 2011;**32**:595-600

[27] Pérez C, Fernandez LE, Sun J, Folch JL, Gill SS, Soberón M, et al. *Bacillus thuringiensis* subsp. *israelensis* Cyt1Aa synergizes Cry11Aa toxin by functioning as a membrane-bound receptor. Proceedings of the National Academy of Sciences. 2005;**102**:18303-18308

[28] Zhang L, Zhao G, Hu X, Liu J, Li M, Batool K. Cry11Aa interacts with the ATP-binding protein from *Culex quinquefasciatus* to improve the toxicity. Journal of Agricultural and Food Chemistry. 2017;**65**:10884-10890

[29] Land M, Miljand M. Biological control of mosquitoes using *Bacillus thuringiensis israelensis*: A pilot study of effects on target organisms, non-target organisms and humans. Mistra Evi EM. 2014;**PS4**

[30] Adang MJ, Crickmore N, Jurat-Fuentes JL. Diversity of *Bacillus thuringiensis* crystal toxins and mechanism of action. Advances in Insect Physiology. 2014;**47**:39-87

[31] Raymond B, Johnston PR, Nielsen LC, Lereclus D, Crickmore N. *Bacillus thuringiensis*: An impotent pathogen? Trends in Microbiology. 2010;**18**:189-194

[32] Sanchis V, Bourguet D. *Bacillus thuringiensis*: Applications in agriculture and insect resistance management. A review. Agronomy for Sustainable Development. 2008;**28**:11-20

[33] Granados B. Founder's Lectures. 29th Annual Meeting of the Society

for Invertebrate Pathology and IIIRD International Colloquium on *Bacillus thuringiensis*. Córdoba, Spain: Society for Invertebrate Pathology; 1996

[34] Galán WLJ. Microbiología y biotecnología avanzada. Ediciones del Instituto de Biotecnología: Monterrey, Nuevo León. 2017;**1**:225-231

[35] Dulmage HT. Insecticidal activity of HD-1, a new isolate of *Bacillus thuringiensis* var. *alesti*. Journal of Invertebrate Pathology. 1970;**15**:232-239

[36] Couch T, Ross D. Production and utilization of *Bacillus thuringiensis*. Biotechnology and Bioengineering. 1980;**22**:1297-1304

[37] Dulmage HT. Assay and standardization of microbial insecticides. Annals of the New York Academy of Sciences. 1973;**217**:187-199

[38] Ponce MEE. Recuperación de cepas HD de *Bacillus thuringiensis*, su propagación en 14 medios de producción y la evaluación de la actividad tóxica contra *Trichoplusia ni* (Háber) y *Heliothis virescens* (Fabricius). [Doctoral dissertation]. San Nicolas, Nuevo León: Universidad Autónoma de Nuevo León; 1996

[39] Dulmage HT. Production of the spore-Sendotoxin complex by variants of *Bacillus thuringiensis* in two fermentation media. Journal of Invertebrate Pathology. 1970;**16**:385-389

[40] Galán-Wong L, Gamiño-Hernández R, Fernández-Chapa D, García-Díaz G, De La Garza-Ramos M, Guajardo-Barbosa C, et al. Persistence of toxic activity of fermentation extracts from *Bacillus thuringiensis var. israelensis* after more than three decades of storage. International Journal of Microbiology. 2017;**1**:1-6

[41] Galán WLJ, Damas BL, Tamez GP, Rodriguez PC, Pereyra AB, Tamez GR, et al. Retrospective view of the contributions of Dr. Howard T. Dulmage to *Bacillus thuringiensis subsp. israelensis* research (memorial lecture). Israel Journal of Entomology. 1998;**32**:15-21

[42] Realpe M, Montoya D, Orduz S. *Bacillus thuringiensis:* Legado para el siglo XXI. Revista Colombiana de Biotecnología. 1998;**1**:11-27

[43] Dulmage HT, McLaughlin RE, Lacey LA, Couch TL, Alls RT, Rose RI. HD-968-S-1983, a proposed US standard for bioassays of preparations of *Bacillus thuringiensis subsp. israelensis-H*-14. Bulletin of the ESA. 1985;**31**:31-34

[44] Dulmage HT, Correa JA, Martinez AJ. Coprecipitation with lactose as a means of recovering the sporecrystal complex of *Bacillus thuringiensis*. Journal of Invertebrate Pathology. 1970;**15**:15-20

[45] Vandekar, M. y H.T. Dulmage. Guidelines for Production of *B. thuringiensis* H-14. 1983. UNDP/World Bank/W.H.O. Geneva, Switzerland

[46] Panwar BS, Ram C, Narula RK, Kaur S. Pool deconvolution approach for high-throughput gene mining from *Bacillus thuringiensis*. Applied Microbiology and Biotechnology. 2018;**102**:1467-1482

[47] Fiuza LM, Polanczyk RA, Crickmore *N. Bacillus thuringiensis and Lysinibacillus sphaericus*: Characterization and Use in the Field of Biocontrol. Springer, Cham, Switzerland. 2017. ISBN: 978-3-319-56678-8

[48] Blackburn MB, Martin PA, Kuhar D, Farrar RR Jr, Gundersen-Rindal DE. Phylogenetic distribution of phenotypic traits in Bacillus thuringiensis determined by multilocus sequence analysis. PLoS One. 2013;**8**:e66061

[49] Reinoso PY, Del Rincón CMC, Ibarra EJ. Characterization of a highly toxic strain of *Bacillus thuringiensis*

serovar *kurstaki* very similar to the HD-73 strain. FEMS Microbiology Reviews. 2016;**363**:1-6

[50] Khorramnejad A, Talaei HR, Hosseininaveh V, Bel Y, Escriche B. Characterization of new *Bacillus thuringiensis* strains from Iran, based on cytocidal and insecticidal activity, proteomic analysis and gene content. BioControl. 2018;**63**:807-818

[51] Armada E, Azcon R, Lopez-Castillo OM, Calvo-Polanco M, Ruiz Lozano JM. Autochthonous arbuscular mycorrhizal fungi and *Bacillus thuringiensis* from a degraded Mediterranean area can be used to improve physiological traits and performance of a plant of agronomic interest under drought conditions. Plant Physiology and Biochemistry. 2015;**90**:64-74

[52] Aceves-Diez AE, Estrada-Castañeda KJ, Castañeda-Sandoval LM. Use of *Bacillus thuringiensis* supernatant from a fermentation process to improve bioremediation of chlorpyrifos in contaminated soils. Journal of Environmental Management. 2015;**157**:213-219

[53] Periyasamy A, Kkani P, Chandrasekaran B, Ponnusamy S, Viswanathan S, Selvanayagam P, et al. Screening and characterization of a non-insecticidal *Bacillus thuringiensis* strain producing parasporal protein with selective toxicity against human colon cancer cell lines. Annals of Microbiology. 2016;**1**:1-12

[54] Singh M, Kumar P, Patel SK, Kalia VC. Production of polyhydroxyalkanoate co-polymer by *Bacillus thuringiensis*. Indian Journal of Microbiology. 2013;**53**:77-83

[55] Gutiérrez-Chávez AJ, Martínez-Ortega EA, Valencia-Posadas M, León Galván MF, de la Fuente-Salcido NM, Bideshi DK, et al. Potential use of *Bacillus thuringiensis* bacteriocins to control antibiotic-resistant bacteria associated with mastitis in dairy goats. Folia Microbiolgy. 2016;**61**:11-19

[56] ISAAA 2017 Global Status of Commercialized Biotech/GM Crops in 2017: Biotech Crop Adoption Surges as Economic Benefits Accumulate in 22 years (ISAAA Brief No. 53). Ithaca, NY: ISAAA

[57] Roush RT. Two-toxin strategies for management of insecticidal transgenic crops: Can pyramiding succeed where pesticide mixtures have not? Philosophical Transactions of the Royal Society of London B. 1998;**353**:1777-1786

[58] Carrière Y, Crickmore N, Tabashnik BE. Optimizing pyramided transgenic Bt crops for sustainable pest management. Nature Biotechnology. 2015;**33**:161-168

[59] Tabashnik BE, Brévault T, Carrière Y. Insect resistance to Bt crops: Lessons from the first billion acres. Nature Biotechnology. 2013;**31**:510-521

[60] Fujimoto H, Itoh K, Yamamoto M, Kyozuka J, Shimamoto K. Insect resistant rice generated by introduction of a modified-endotoxin gene of *Bacillus thuringiensis*. Nature Biotechnology. 1993;**11**:1151-1155

[61] Chen M, Shelton A, Ye GY. Insect-resistant genetically modified rice in China: From research to commercialization. Annual Review of Entomology. 2011;**56**:81-101

[62] Ye GY, Shu QY, Yao HW, Cui HR, et al. Field evaluation of resistance of transgenic rice containing a synthetic cry1Ab gene from *Bacillus thuringiensis* Berliner to two stem borers. Journal of Economic Entomology. 2001;**94**:271-276

[63] Riaz N, Husnain T, Fatima T, Makhdoom R, Bashir K, et al. Development of Indica Basmati rice harboring two insecticidal genes

for sustainable resistance against lepidopteran insects. South African Journal of Botany. 2006;**72**(2):217-223

[64] High SM, Cohen MB, Shu QY, Altosaar I. Achieving successful deployment of Bt rice. Trends Plant Science. 2004;**9**:286-292

[65] Jian Z, Wang F, Li Z, Chen Y, Ma X, Nie L, et al. Grain yield and nitrogen use efficiency responses to N application in Bt (Cry1Ab/Ac) transgenic two-line hybrid rice. Field Crops Research. 2014;**155**:184-191

[66] Xu C, Cheng J, Lin H, Lin C, Gao J, Shen Z. Characterization of transgenic rice expressing fusion protein Cry1Ab/Vip3A for insect resistance. Scientific reports. 2018;**8**:15788

[67] Li ZY, Sui H, Xu YB, Han LZ, Chen FJ. Effects of insect-resistant transgenic Bt rice with a fused Cry1AbþCry1Ac gene on population dynamics of the stem borers, *Chilo suppressalis* and *Sesamia inferens*, occurring in paddyfield. Acta Ecologica Sinica. 2012;**32**:1783-1789

[68] Wang Z, Fang L, Zhou Z, Pacheco S, Gómez I, Song F, et al. Specific binding between *Bacillus thuringiensis* Cry9Aa and Vip3Aa toxins synergizes their toxicity against Asiatic rice borer (Chilo suppressalis). Journal of Biological Chemistry. 2018;**293**:11447-11458

[69] Saito T, Hirai K, Way MO. The rice water weevil, *Lissorhoptrus oryzophilus* Kuschel (Coleoptera: Curculionidae). Applied Entomology and Zoology. 2005;**40**:31-39

[70] Lupi D, Cenghialta C, Colombo M. Adult feeding by the rice water weevil *Lissorhoptrus oryzophilus* on different host plants. Bulletin of Insectology. 2009;**62**:229-236

[71] Tindall KV, Stout MJ. Use of common weeds of rice as hosts for the rice water weevil (Coleoptera: Curculionidae). Environmental Entomology. 2003;**32**:1227-1233

[72] Soderlund DM, Clark JM, Sheets LP, Mullin LS, Piccirillo VJ, Sargent D, et al. Mechanisms of pyrethroid neurotoxicity: Implications for cumulative risk assessment. Toxicology. 2002;**171**:3-59

[73] Aghaee M-A, Godfrey LD. The efficacy of *Bacillus thuringiensis spp. galleriae* against rice water weevil (Coleoptera: Curculionidae) for integrated pest management in california rice. Journal of Economic Entomology. 2015;**108**(1):45-52

[74] Fernández C, Juncosa R. Biopesticidas: La agricultura del futuro. Phytoma. 2002;**141**:14-19

[75] Ferré J, Van Rie J. Biochemistry and genetics of insect resistance to *Bacillus thuringiensis*. Annual Review of Entomology. 2002;**47**:501-533

[76] Tabashnik BE, Van Rensburg JBJ, Carrière Y. Field-evolved insect resistance to Bt crops: Definition, theory, and data. Journal of Economic Entomology. 2009;**102**:2011-2025

[77] Jakka S, Shrestha R, Gassmann A. Broad-spectrum resistance to *Bacillus thuringiensis* toxins by western corn rootworm (*Diabrotica virgifera virgifera*). Scientific Reports. 2016;**6**:1

[78] Bravo A, Likitvivatanavong S, Gill SS, Soberón M. *Bacillus thuringiensis*: A story of a successful bioinsecticide. Insect Biochemistry and Molecular Biology. 2011;**41**:423-431

[79] Tikar S, Prakash S. Fly ash-based *Bacillus thuringiensis israelensis* formulation: An ecofriendly approach. The Indian Journal of Medical Research. 2017;**146**(**6**):680

[80] Nava PE, García GC, Camacho BJR, Vázquez MEL. Bioplaguicidas: Una

opción para el control biológico de plagas. Ra Ximhai. 2012;**8**:17-29

[81] Tadros F. Applied Surfactants, Principles and Applications. Weinheim: Wiley-VCH Verlag GmbH and Co. KGaA; 2005. pp. 187-256

[82] Lyn ME, Burnett D, Garcia AR, Gray R. Interaction of water with three granular biopesticide formulations. Journal of Agricultural and Food Chemistry. 2010;**58**:1804-1814

[83] Brar SK, Verma M, Tyagi RD, Valero JR. Recent advances in downstream processing and formulations of *Bacillus thuringiensis* based biopesticides. Process Biochemistry. 2006;**41**:323-342

[84] Aneja KR, Khan SA, Aneja A. Biopesticides an eco-friendly pestmanagement approach in agriculture: Status and prospects. Kavaka. 2016;**47**:145-154

[85] Kumar S, Singh A. Biopesticides: Present status and the future prospects. Journal of Fertilizers & Pesticides. 2015;**6**:129

[86] Portela DDD, Chaparro GA, López PSA. La biotecnología de *Bacillus thuringiensis* en la agricultura. Nova. 2013;**11**:87-96

[87] Tamez GP, Galán WLJ, Medrano RH, García GC, Rodríguez PC, Gómez FRA, et al. Bioinsecticidas: Su empleo, producción y comercialización en México. Ciencia UANL. 2001;**4**:143-152

[88] Sauka DH. *Bacillus thuringiensis*: ¿nuevas aplicaciones para un viejo conocido? Revista Argentina de Microbiologia. 2017;**49**:123-124

[89] Ruan L, Crickmore N, Peng D, Sun M. Are nematodes a missing link in the confounded ecology of the entomopathogen *Bacillus thuringiensis*? Current Trends in Microbiology. 2015;**23**:341-346

[90] Dunstand GE, Peña CG, Hallal CC, Pérez MM, Hernández VVM, Morales MJ, et al. Acaricidal effect and histological damage induced by *Bacillus thuringiensis* protein extracts on the mite Psoroptes cuniculi. Parasites & Vectors. 2015;**8**:1

[91] Shrestha A, Sultana R, Chae JC, Kim K, Lee KJ. *Bacillus thuringiensis* C25 which is rich in cell wall degrading enzymes efficiently controls lettuce drop caused by *Sclerotinia minor*. European Journal of Plant Pathology. 2015;**142**:577-589

[92] Raddadi N, Cherif A, Ouzari H, Marzorati M, Brusetti L, Boudabous A, et al. *Bacillus thuringiensis* beyond insect biocontrol: Plant growth promotion and biosafety of polyvalent strains. Annals of Microbiology. 2007;**57**:481-494

[93] De la Fuente SNM, Casados VLE, Barboza CJE. Bacteriocins of *Bacillus thuringiensis* can expand the potential of this bacterium to other areas rather than limit its use only as microbial insecticide. Canadian Journal of Microbiology. 2013;**59**:515-522

[94] Ondarza BMA. Biopesticidas: Tipos y aplicaciones en el control de plagas agrícolas. Agroproductividad. 2017;**10**:31-36

[95] Damalas CA, Koutroubas SD. Current status and recent developments in biopesticide use. Agriculture. 2018;**8**:1-6